INTRODUCTION TO STATICS

PRENTICE-HALL

SERIES IN ENGINEERING SCIENCE

PRENTICE-HALL, INTERNATIONAL, INC., *London*
PRENTICE-HALL OF AUSTRALIA, PTY, LTD., *Sydney*
PRENTICE-HALL OF CANADA, LTD., *Toronto*
PRENTICE-HALL OF INDIA PRIVATE LTD., *New Delhi*
PRENTICE-HALL OF JAPAN, INC., *Tokyo*

INTRODUCTION TO STATICS

IRVING H. SHAMES

Faculty Professor
of Engineering and Applied Science
State University of New York at Buffalo

PRENTICE-HALL, INC., ENGLEWOOD CLIFFS, N. J.

PREFACE

This text is intended to provide a short and efficient, but nevertheless, reasonably self-contained treatment of the key aspects of the statics of particles and rigid bodies. The prime source of material for this book is the author's *Engineering Mechanics: Statics*. I have reviewed this earlier text carefully, sifting out the salient features for developing what I consider a minimal, yet cogent and meaningful introduction to the subject. To accomplish this, I have tried to make as much use as possible of earlier course work in physics while, at the same time, not becoming overly dependent on this contribution. For instance, in Chapter 1, I have assumed that the student understands such material as units, dimensions, dimensional homogeneity, gravitational law, etc. However, most of these topics are presented as review problems at the end of the chapter coupled with sufficient reminders to bring this material back into focus. As for vector algebra, I have included in the appendix a self-teaching presentation whereby a student with some exposure to vectors in physics and mathematics can proceed on his own to recall this material and enhance his skill in this area.

In a normal three-credit course, this text can be covered effectively, I believe, in five to eight weeks. At Buffalo, lecturing to the entire sophomore class (up to 300 students), I cover the entire text in five weeks in a four-credit course—and expect to reduce this to four weeks in the future.

This leaves time to proceed with a study of dynamics or of strength of materials.

The contents of this text can be explained as follows. In Chapter 1, the primary effort (once certain idealizations of mechanics have been discussed), sets forth the concept of equivalence between vectors. In Chapter 2, we first discuss the position vector, the moment of a force about a point and a line, and the couple and couple-moment. We are then ready to explore the concept of equivalence between force systems acting on rigid bodies. We accordingly consider a translation of a force to a parallel position, allowing us to develop the idea of the resultant force system. Having done this, we consider the simplest resultant for coplanar and parallel force systems as well as distributed force systems. This sets the stage for the consideration of the equations of equilibrium in Chapter 3. After careful delineation of the free-body diagram, we reach our study of equations of equilibrium by reasoning how to eliminate the simplest resultant for any particular force system acting on a free body. Many examples are presented to illustrate this procedure. Included in this chapter is a discussion of simple trusses and of friction as it pertains to rigid-body mechanics.

I wish to thank my colleagues at State University of New York at Buffalo, who have assisted in the development of this shorter approach to statics. My association with them has been most helpful in the preparation of this text.

IRVING H. SHAMES

Buffalo, N.Y.

CONTENTS

INTRODUCTION TO STATICS

Chapter 1

INTRODUCTORY CONCEPTS

1.1. Introduction

Mechanics is that branch of physics having to do with mechanical effects (as opposed to chemical and thermal effects) on bodies. As such, mechanics forms the basis of much engineering analysis and design. Also, it is one of the oldest sciences, with a history extending back to the time of the Greek and Roman civilizations. However, mechanics as we know it today was set forth by Sir Isaac Newton in 1687 when he presented his famous three laws. Although the passage of time has seen limitations placed on the range of validity of Newtonian mechanics, it nevertheless has held up exceedingly well as a science for over three centuries. One of the limitations referred to is the fact that Newtonian mechanics becomes invalid for atomic behavior, i.e., it does not apply to bodies of atomic size acting within atomic distances. Here the fundamental equation is Schrödinger's equation rather than Newton's laws, and we become involved in quantum mechanics rather than Newtonian mechanics. Also, when the speed of a body approaches the speed of light, many of the concepts underlying Newtonian mechanics must be broadened in order to accurately portray vital effects at these speeds. As a result we become involved with rel-

1

ativistic mechanics. Although quantum and relativistic mechanics continually grow in importance for engineers, there is no less need to develop a firm grasp of Newtonian mechanics both for important direct use in engineering and applied science and for forming a firm basis for the study of the aforementioned newer sciences.

In this text we shall present a compact but reasonably complete introduction to a part of mechanics called the statics of particles and rigid bodies (both particles and rigid bodies will be discussed in Section 1.2). The dynamics of these bodies is left for another volume.* We shall later set forth precisely what we mean by *statics*.

1.2. Idealizations of Mechanics

To be able to represent an action using the known laws of physics and also to be able to form equations simple enough to be susceptible to mathematical computational techniques, invariably in our deliberations we must replace the actual physical action, and the participating bodies, with hypothetical, highly simplified substitutes. We must be sure, of course, that the results of our substitutions have some reasonable correlation with reality. All analytical physical sciences must resort to this technique, and, consequently, their computations are not cut and dried but involve a considerable amount of imagination, ingenuity, and insight into physical behavior. We shall at this time set forth the most fundamental idealizations of mechanics and a bit of the philosophy involved in scientific analysis.

The Continuum

Even the simplification of matter into molecules, atoms, electrons, etc., is too complex a picture for many problems of engineering mechanics. In most problems, we are interested only in the average measurable manifestations of these elementary bodies. Pressure, density, and temperature are actually the gross effects of the actions of the many molecules and atoms, and they can be conveniently assumed to arise from a hypothetically continuous distribution of matter, which we shall call the *continuum*, instead of from a conglomeration of discrete bodies. Without such an artifice, we would have to consider the action of each of these elementary bodies—a virtual impossibility for most problems.

---

*See the author's text *Engineering Mechanics, Volume II: Dynamics* (Englewood Cliffs, N.J.: Prentice-Hall, Inc., 1966).

The Rigid Body

In many cases involving the action on a body by a force, we simplify the continuum concept even further. The most elemental case is that of a rigid body, which is a continuum that undergoes theoretically no deformation whatever. Actually every body must deform a certain amount, but in many cases the deformation is too small to affect the desired analysis. It is then preferable to consider the body as rigid and proceed with the simplified computations. For example, assume we are to determine the forces transmitted by a beam to the earth as the result of some load P (Fig. 1.1). If P is reasonably small, the beam will undergo little deflection and we can carry out a straightforward simple analysis as if the body were indeed rigid. If we were to attempt a more highly

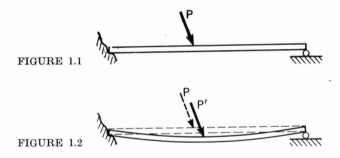

FIGURE 1.1

FIGURE 1.2

accurate analysis—even though a slight increase in accuracy is not required—we would then need to know the position that the load assumes relative to the earth *after* the beam has ceased to deform, as shown in an exaggerated manner in Fig. 1.2. This then is a very difficult task, especially when we consider that the support must also "give" in a certain way. Although the alternative to a rigid-body analysis here leads us to a needlessly difficult calculation, situations nevertheless do arise in which more realistic models must be employed to yield the required accuracy. *The guiding principle is to make such simplifications as are consistent with the required accuracy of the results.*

The Particle

The particle is defined as an object that has no size but that has a mass. Perhaps this doesn't sound like a very helpful definition for engineers to employ, but it is actually one of the most useful in mechanics. In computing the trajectory of a planet, for example, it is the mass of the planet and not its size that is significant. Hence, we can consider planets as particles for such computations. On the other hand, take a figure skater spinning on the ice, whose revolutions are controlled so beautifully by the orientation of the body. In this motion, the distribution of the body is significant, and since a particle, by definition, can have no distribution, it is patently clear that a particle cannot represent the skater in this case. If, however, the skater should be billed as the "human cannonball on skates" and be shot out of a large gun, then it would be possible to consider him as a single particle in ascertaining his trajectory, since arm and leg movements that were significant while he was spinning on the ice would have little effect on the arc traversed by the main portion of his body.

Point Force

A finite force exerted on one body by another must cause a finite amount of local deformation and always creates a finite area of contact between the bodies through which the force is transmitted. However, since we have formulated the concept of the rigid body, we should also be able to imagine that a finite force is transmitted through an infinitesimal area or point. This simplification of a force distribution is called a *point force*. In the many cases where the actual area of contact in a problem is very small but is not known exactly, the use of the concept of the point force results in little sacrifice in accuracy. In Figs 1.1 and 1.2 we actually employed the graphical representation of the point force.

Many other simplifications pervade mechanics. The perfectly elastic body, the frictionless fluid, etc., will become quite familiar to you as you study various phases of mechanics.

1.3. Vector and Scalar Quantities

Certain quantities require only a single value for their complete specification. (Temperature and pressure are two such quantities.) We call these quantities *scalars* to distinguish them from other quantities having additional requirements for their full specification. As an illustration of the latter, note that to specify fully the velocity of a particle we must give:

a. *The value* or *magnitude* of the velocity by stating the number of scale units (feet per second, centimeters per second, etc.).

b. The *direction** of the velocity relative to some convenient reference.

To represent the velocity and other such quantities graphically, we may use an arrow; the length of the arrow then conveys the value or magnitude of the velocity. You will note we have already used such arrows in Figs. 1.1 and 1.2.

Now, certain of the quantities having magnitude and direction, such as velocity and force, combine their effects in a special way. Thus, the combined effect of two forces acting on a particle, as shown in Fig. 1.3, corresponds to a single force that may be shown by experiment to be equal to the diagonal of a parallelogram formed by the graphical representation of the forces. That is, the quantities add according to the *parallelogram law*. All quantities that have magni-

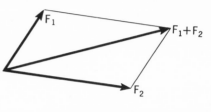

FIGURE 1.3

tude and direction and that add according to the parallelogram law are called *vector quantities*. A vector quantity will be denoted with a boldface letter, which in the case of force becomes **F**.†

The reader may ask: don't all quantities having magnitude and direction combine according to the parallelogram law and therefore become vector quantities? No, not all of them do. One very important example will be pointed out after we reconsider Fig. 1.3. In the construction of the parallelogram it matters not which force is laid out first. In other words, "F_1 combined with F_2" gives the same result as "F_2 combined with F_1." In short, the combination is *commutative*. If a combination is not commutative, then it cannot in general be represented by a parallelogram operation and is thus not a vector. With this in mind, consider the angle of rotation of a body about some axis. We can associate a magnitude (degrees or radians) and a direction (the axis and a stipulation of clockwise or counterclockwise) with this quantity. However, the angle of rotation cannot be considered a vector because, in general, two rotations about different axes cannot be replaced by a single rotation consistent with the parallelogram law. The easiest way to show this is to demonstrate that the combination of rotations is not

---

*The direction as used here includes the specification of the orientation of the line of action of the velocity as well as the sense of the velocity. The line of action and sense are as defined in the students' earlier physics courses.

†We also represent vectors in this text with a superscript arrow. Thus **F** and $\vec{F}$ are alternate ways to represent the force vector.

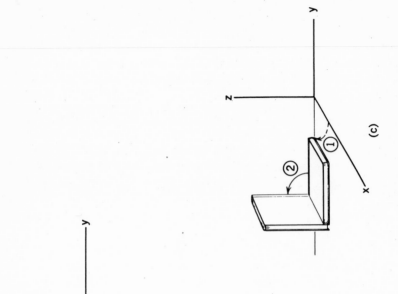

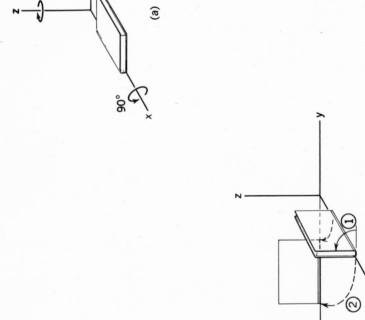

FIGURE 1.4

commutative. In Fig. 1.4 (a) a book is to be given two rotations—a 90° rotation counterclockwise about the x axis and a 90° clockwise rotation about the z axis looking in toward the origin in both cases. In Fig. 1.4(c) the sequence of combination is reversed from that in Fig. 1.4(b) and you can see how it alters the final orientation of the book. Angular rotation, therefore, is not a vector quantity, since the parallelogram law is not valid for combinations of angular rotations.*

1.4. Equality and Equivalence of Vectors

We shall avoid many pitfalls in the study of mechanics if we clearly make a distinction between the equality and the equivalence of vectors.

Two vectors are equal if they have the same dimensions, magnitude, and direction. In Fig. 1.5 the velocity vectors of three particles have equal length, are identically inclined toward the reference xyz, and have the same sense. Although they have different lines of action, they are nevertheless equal according to the definition.

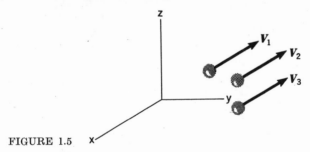

FIGURE 1.5

Two vectors are equivalent in a certain capacity if each produces the very same effect in this capacity. If the criterion in Fig. 1.5 is change of elevation of the particles or total distance traveled by the particles, all three vectors give the same result. They are, in addition to being equal, also equivalent for these capacities. If the absolute height of the particles above the xy plane is the question in point, these vectors will not be equivalent despite their equality. Thus, it must be emphasized that *equal vectors need not always be equivalent*; *it depends entirely on the situation at hand.* Furthermore, vectors that are not equal may still be

---

*However, *vanishingly small* rotations can be considered as vectors since the commutative law applies for the combination of such rotations. This is an important consideration when discussing the angular velocity vector in dynamics.

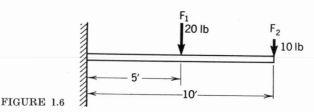

FIGURE 1.6

equivalent in some capacity. Thus, in the beam in Fig. 1.6 forces F_1 and F_2 are unequal since their magnitudes are 10 lb and 20 lb respectively. However, it is clear from elementary physics that their moments about the base of the beam are equal and so the forces have the same "twisting" action at the fixed end of the beam. In that capacity, the forces are equivalent. If, however, we are interested in the deflection of the free end of the beam resulting from each force, there is no longer an equivalence between the forces, since each will give a different deflection.

To sum up, the *equality* of two vectors is determined by the vectors themselves, and the *equivalence* between two vectors is determined by the situation at hand.*

In problems of mechanics, we can profitably delineate three classes of situations concerning equivalence of vectors:

  a. *Situations in which vectors may be positioned anywhere in space without loss or change of meaning provided magnitude and direction are kept intact.* Under such circumstances the vectors are called *free vectors.* For example, the velocity vectors in Fig. 1.5 are free vectors as far as total distance traveled is concerned.†

  b. *Situations in which vectors may be moved along a line collinear with the vector itself without loss or change of meaning provided magnitude and sense are kept intact.* Under such circumstances the vectors are called *transmissible vectors.* For example, in towing the object in Fig. 1.7, we may apply the force anywhere along the rope AB or may push at point C. The resulting motion is the same in all cases, so the force is a transmissible vector for this purpose.

  c. *Situations in which the vectors must be applied at definite points.* The point may be represented as the tail or head of the arrow in the graphical representation. For this case no other position of applica-

---

*A basic equivalence, however, that is always valid in mechanics exists between the sum of concurrent vectors and the system of component vectors.

†In Chapter 2 another basic equivalence will be set forth when it is shown that the moment of a couple is always a free vector.

tion leads to equivalence. Under such circumstances the vector is called a *bound vector*. For example, if we are interested in the deformation induced by forces in the body in Fig. 1.7, we must be more selective in our actions than we were when all we wanted to know was the motion of the body. Clearly force F will cause a different deformation when applied at point C than it will when applied to point A. The force is thus a bound vector for this consideration.

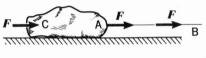

FIGURE 1.7

We shall be concerned throughout this text with considerations of equivalence of vectors.

1.5. Laws of Mechanics

The entire structure of mechanics rests on relatively few basic laws. Nevertheless, for the student to comprehend these laws sufficiently to undertake novel and varied problems much study will be required. We shall now discuss briefly the following laws that are considered to be the foundation of mechanics:

a. Newton's first and second laws of motion.
b. Newton's third law.

a. Newton's First and Second Laws of Motion

These laws were first stated by Newton in roughly these words:

Every body continues in a state of rest or uniform motion in a straight line unless it is compelled to change that state by forces imposed on it. The change of motion is proportional to the natural force impressed and is made in a direction of the straight line in which the force is impressed.

Notice that the words "rest," "uniform motion," and "change of motion" appear in the above statements. For such information to be meaningful we must have some frame of reference relative to which these states of motion can be described. We may then ask: relative to what reference in space does every body remain at "rest or move uniformly along a straight line" in the absence of any forces?* Or, in the case of a force acting on the body, relative to what reference in space is the "change in motion proportional to the force"? Experiment indicates that the so-called fixed stars act as a reference for which the first and second laws of Newton are highly accurate. Later we will se that any other system that moves uniformly and without rotation relative to the fixed stars may be used as a reference with equal accuracy. All such references are called *inertial references*. The earth's surface is usually employed as a reference in engineering work. Because of the rotation of the earth and the variations in its motion around the sun, it is not strictly speaking an inertial reference. However, the departure is so small for most situations (an exception is the motion of long-range rockets) that the error incurred is very slight. We shall, therefore, usually consider the earth's surface as an inertial reference, but will keep in mind the somewhat approximate nature of this step.

As a result of the preceding discussion, we may define *equilibrium as that state of a body in which it is at rest relative to an inertial reference or where each point of the body moves uniformly along straight parallel lines relative to an inertial reference.* The converse of Newton's first law then stipulates for the equilibrium state that there must be zero net force (or equivalent action of zero net force) acting on the body. Many situations fall into this category. The study of bodies in equilibrium is called *statics*. The statics of particles and rigid bodies will be the prime consideration in this text.

b. Newton's Third Law

Newton stated in his third law:

To every action there is always opposed an equal reaction, or the mutual actions of two bodies upon each other are always equal and directed to contrary points.

This is illustrated graphically in Fig. 1.8, where the action and reaction between two bodies arise from direct contact. Other important actions in which Newton's third law holds are gravitational attractions and electrostatic forces between charged particles. It should be pointed

---

*Some authors (for example, A. Sommerfeld, *Mechanics*, Academic Press) consider the first law as the definition of a reference in space for which the second law is then valid.

Date Due Slip

A-B Tech Community College
09/20/16 01:18PM

* * * * * * * * * * * * * * * * * *

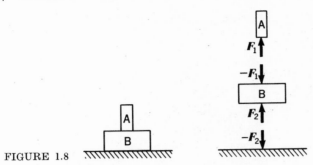

FIGURE 1.8

out that there are actions that do not follow this law, notably the electromagnetic forces between charged moving bodies.*

1.6. Closure

In this introductory chapter we have presented some of the idealizations of mechanics that are of immediate use. Other idealizations will be presented as needed. With these idealizations we can simplify many problems to a state where they may be described by reasonably simple mathematical formulations. In doing so we become involved with scalar and vector quantities. We dwelled on the important question of the equality and equivalence of vectors—a consideration that will be the key idea toward arriving at the equations of equilibrium later in the text. Since we shall consider only the state of

Table A. COMMON SYSTEMS OF UNITS

C.G.S.		M.K.S.	
Mass	Gram	Mass	Kilogram
Length	Centimeter	Length	Meter
Time	Second	Time	Second
Force	Dyne	Force	Newton
English		American practice	
Mass	Pound Mass	Mass	Slug or Pound Mass
Length	Foot	Length	Foot
Time	Second	Time	Second
Force	Poundal	Force	Pound Force

---

*Electromagnetic forces between charged moving particles are equal and opposite but are not collinear and hence are not "directed to contrary points."

Table B. EQUIVALENCE RELATIONS BETWEEN UNITS

1 in. ≡ 2.54 cm	1 slug ≡ 32.2 lbm
1 ft ≡ 30.5 cm	1 gram ≡ 2.205×10^{-3} lbm
1 ft ≡ 0.305 meter	1 gram ≡ 0.685×10^{-4} slug
5280 ft ≡ 1 mile	

1 lbf ≡ 445,000 dynes
1 lbf ≡ 32.2 poundals
1 lbf ≡ 16 ounces
1 Newton ≡ 10^5 dynes

equilibrium (defined as a state of rest or uniform motion of all points relative to an inertial reference) in the text, these considerations will prove critical in much of this text.

In the problems following this chapter the student will have the opportunity to review some elementary aspects of mechanics that were presented in his earlier physics courses. Tables A and B of units and their equivalents will be useful in this regard.

Problems

▶ The basic dimensions for mechanics are usually taken as length, mass, and time, where L, M, and t are, respectively, the dimensional representation of these quantities. Other quantities are developed from these concepts. Velocity is such a quantity and it has a dimensional representation given as (L/t^2).

1. (a) Express density dimensionally. (b) How many scale units of density in the metric system of centimeters, grams, and seconds are equivalent to a scale unit in the American system using (1) slugs, feet, and seconds? (2) lbm, ft, sec?

2. What are the dimensional representations of the following quantities: (a) acceleration, (b) pressure, (c) angle of rotation, (d) specific volume?

3. The escape velocities for the earth and the moon are known to be 11.2 km/sec and 2.4 km/sec, respectively. Determine these escape velocities in units of miles/hour.

▶ Equations of physics must be dimensionally homogeneous; i.e., each grouping in the equation must have the same dimensional representation. Otherwise the equation is valid only for one system of units and hence cannot represent a physical process, which, of course, knows nothing of man-made units.

4. Explain why an equation that is not dimensionally homogeneous cannot be valid for all systems of units. Demonstrate this.

5. The Newton viscosity law says that the frictional resistance, τ, in a fluid, given as force per unit area, is proportional to the distance rate of change of velocity dV/dy. The proportionality constant, μ, is called the *coefficient of viscosity*. (a) What dimensions must it have? (b) What is the relation between the scale units for μ in the English system (slug, ft, sec) and in the metric system (gm, cm, sec)?

▶ A group of terms formed by products and divisors is termed dimensionless if the dimensional representation of the terms can be completely canceled. Such quantities are particularly important in fluid mechanics and heat transfer.

13

6. The resistance of a body moving through a fluid, such as a rocket moving through air, is sometimes expressed by the following equation:

$$F = \tfrac{1}{2} C_D \rho V^2 A$$

where F is the resistance
$\quad\quad C_D$ is the coefficient of drag
$\quad\quad \rho$ is the density of the fluid
$\quad\quad V$ is the velocity of the object relative to the undisturbed fluid
$\quad\quad A$ is the cross-sectional area of the body at right angles to the motion

What is the dimensional representation of C_D?

7. In the study of the motion of boats a grouping of terms called the *Froude* number is important. A form of the Froude number, Fr, can be given as:

$$\mathrm{Fr} \equiv \frac{\rho V^2}{L\gamma}$$

where ρ is the density of the water, V is the speed of the boat, L is a characteristic length of the boat, and γ is the specific weight (weight per unit volume) of the water. Show that the Froude number is dimensionless. Show also that the Froude number can be given as V^2/Lg.

8. A group of terms useful in the study of the surface effects of a liquid on a floating object moving on it is the so-called *Weber* number, W_e, given as:

$$W_e \equiv \frac{\rho V^2 L}{\sigma}$$

where
$\quad\quad \rho$ is the density of the liquid
$\quad\quad \sigma$ is the surface tension of the liquid
$\quad\quad L$ is a characteristic length of the body
$\quad\quad V$ is the speed of the body.

Like the Froude number of the previous problem, the Weber number is dimensionless. What are the dimensions of σ, the surface tension?

9. The following equation is valid for certain pipe-flow analyses:

$$\frac{\Delta p}{\rho V^2} = f\left[\left(\frac{\rho V D}{\mu}\right), \left(\frac{L}{D}\right)\right]$$

where

    Δp is the drop in pressure along a pipe

    ρ is the density of the fluid

    V is the average velocity of flow

    D is the inside diameter of the pipe

    L is the length over which the pressure drop is measured

    μ is the viscosity of the fluid

The right side of the equation is to be interpreted as some function f of the variables $\rho VD/\mu$ and L/D. Show that such an equation is dimensionally homogeneous for any function f. ($\rho VD/\mu$ is called the *Reynolds number*.) See Prob. 5 for information concerning μ.

10. A pound mass is an amount of matter attracted toward the earth at a specific location on the earth's surface with a force of 1 lb. A slug is the amount of matter that accelerates 1 ft/sec² under the action of a 1-lb force. Define a gram. How many grams are equivalent to a pound mass?

11. In using as your basic system of dimensions mass, length, and time, explain how you could set forth two entirely different ways of measuring force. How would you then ascertain the mutual physical equivalence of the force units from these measurements?

12. Suppose you desired to use velocity, time, and force as a basic system of dimensions. What would then be the dimensional representation for quantities such as length, mass, and acceleration in terms of these dimensions?

13. Engineers on the Continent use kilograms as a measure of mass. However, they also consider kilograms as a measure of force. What do you think a kilogram of force means, and what reservations would you caution in its use?

14. Consider the earth's atmosphere. When and why would you feel obliged to drop the concept of the continuum?

15. A rocket is shot from the earth, with a high rate of spin to maintain its stability (like a football). A relay must close during part of the flight in a certain time. This means that an element of the relay must be given a certain acceleration $\boldsymbol{a}_r$ relative to the rocket. Can we use Newton's law in the form $\boldsymbol{F} = m\boldsymbol{a}_r$ to determine the required force? If not, why not, and what must we do?

16. What is the difference between relativistic restrictions and quantum restrictions on Newton's law?

▶The law of gravitational attraction between bodies of mass M_1 and M_2 is given as:

$$F = G \frac{M_1 M_2}{R^2}$$

where R is the distance between the centers of mass of the bodies (for a uniform sphere the center of mass is the geometric center) and G is the universal gravitational constant.

17. A planet moves near a fixed star. Show that the motion of the planet is independent of its mass.

18. (a) If you were to kick a cannonball on the earth and then the same ball on the moon, would it hurt more on the earth, on the moon, or both the same? (b) If you were to drop the cannonball onto your toe from a certain height, would it hurt more if you did this on the earth or on the moon?

19. The weight of the first Vanguard satellite on the earth was said to be $3\frac{1}{4}$ lb. At the extreme position in its orbit, it was approximately 2500 miles from the earth's surface. What is the weight of the satellite there?

20. In the previous problem, what is the acceleration of gravity to the earth of the Vanguard satellite at its extreme position?

21. The diameter of the moon is 2160 miles. The acceleration of gravity at the surface is 5.32 ft/sec². What is the ratio of the masses of the earth and moon?

22. During the flight of Apollo II to the moon, the center of the earth was approximately 219,000 miles from the center of the moon. At what distance from the earth's center might Apollo II have zero weight?

23. Show that the acceleration of gravity, g, of a body at the earth's surface is given by the formula:

$$g = \frac{GM}{R^2}$$

where G is the universal gravitational constant, M is the mass of the earth, and R is the radius of the earth.

24. In physics handbooks the universal gravitational constant G is given as:

$$G = 6.66 \times 10^{-8} \text{ cm}^3\text{-sec}^{-2}\text{-grams}^{-1}$$

Using the results of Prob. 23, estimate the mass of the earth in tons mass. The radius of the earth may be taken as 4000 miles.

25. Coulomb's law for point charges has a form almost identical to Newton's gravitational law. Thus we have:

$$F = \frac{1}{4\pi\epsilon_0} \frac{q_1 q_2}{r^2}$$

where F is the force between the charges in newtons, q_1, and q_2 are the values of charge in coulombs, r is the distance between charges in meters, and ϵ_0 is a constant called the *permittivity constant* with the value:

$$\epsilon_0 = 8.85 \times 10^{-12} \text{ coulomb}^2/\text{newton-meters}^2$$

(We are here using the M.K.S. system of units.) Compare the electrostatic force between two electrons at a distance of 1 angstrom apart (10^{-8} cm) with the gravitational attraction between these electrons. The mass of an electron is 9.1×10^{-31} kilograms and its charge is 1.6×10^{-19} coulombs. Take the value of G as given by Prob. 24.

Chapter 2

FORCE SYSTEMS

2.1. Position Vector

In this chapter we shall employ a number of useful vector quantities. One such vector is the *position vector*. The position vector r of point P is the directed line segment between some reference point, which is taken as the origin of a coordinate system, and point P (Fig. 2.1). The magnitude of the position vector is the distance between the points. The scalar components of a position vector are simply the coordinates of the point P. To express r in Cartesian components, we then have:

$$r = xi + yj + zk \qquad \qquad \textbf{2.1}$$

A vector going from any point (1) in space to any point (2) is called a *displacement vector* from point (1) to point (2) and is often denoted as ρ. It should be clear that we can express ρ in terms of position vectors and coordinates of points (1) and (2) in the following manner:

$$\rho = r_2 - r_1 = (x_2 - x_1)i + (y_2 - y_1)j + (z_2 - z_1)k \qquad \textbf{2.2}$$

19

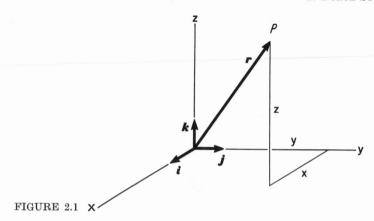

FIGURE 2.1

2.2. Moment of a Force about a Point

In physics, you will recall, the moment of a force F about a point O (see Fig. 2.2) is a vector M whose magnitude equals the product of the force magnitude times the perpendicular distance d from the point to the line of action of the force. And the direction of this vector is perpendicular to the plane of point and force with a sense determined from the familiar right-hand screw rule. In Fig. 2.2 the line of action M is taken through point O.

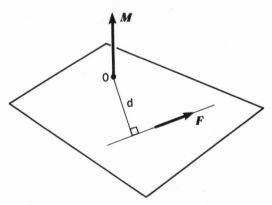

FIGURE 2.2

Another approach is to employ a position vector r from point O to *any point* P along the line of action of force F as shown in Fig. 2.3. The moment of F about point O is then defined as:

$$M = r \times F$$

2.3

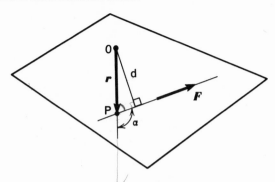

FIGURE 2.3

Clearly we get the same magnitude for M as earlier since $|\mathbf{r} \times \boldsymbol{F}| = rF \sin \alpha = Fd$. Furthermore, both elementary and vector definitions give the same direction for M. Thus, the former stipulates perpendicularity of M to the plane of O and $\boldsymbol{F}$, while the latter stipulates perpendicularity of M to the plane of $\boldsymbol{r}$ and $\boldsymbol{F}$. Clearly these give identical lines of action. And, both have senses consistent with the right-hand screw rule. Thus the elementary definition and the cross-product definition give the same results.

Assuming a reference x, y, z at O, we obtain for M:

$$\boldsymbol{M} = (F_z y - F_y z)\boldsymbol{i} + (F_x z - F_z x)\boldsymbol{j} + (F_y x - F_x y)\boldsymbol{k} \qquad \textbf{2.4}$$

The scalar rectangular components of M are then:

$$M_x = F_z y - F_y z \qquad \textbf{(a)}$$
$$M_y = F_x z - F_z x \qquad \textbf{(b)} \quad \textbf{2.5}$$
$$M_z = F_y x - F_x y \qquad \textbf{(c)}$$

It should be clear that Eqs. 2.4 and 2.5 give the moment M of force $\boldsymbol{F}$ about the origin of reference xyz.

As a final note, it should be apparent that, because we can choose $\boldsymbol{r}$ so as to terminate anywhere along the line of action of $\boldsymbol{F}$ in computing M, we are in effect stipulating that $\boldsymbol{F}$ is a transmissible vector in the computation of M.

EXAMPLE 2.1

We wish to compute the moment of the 1000-lb force about point E in Fig. 2.4. The simplest displacement vector from E to the line of action of the force is obviously $\overrightarrow{EA}$ and so we can say:

$$\boldsymbol{\rho}_{EA} = -20\boldsymbol{j}$$

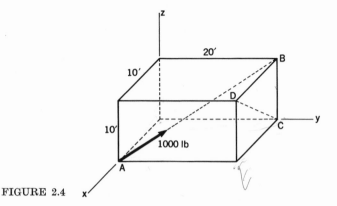

FIGURE 2.4

Next we wish to ascertain the force in terms of rectangular components. We may do this by employing a unit vector from A to B, which we denote as $\hat{t}_{AB}$ so that:

$$F = 1000\hat{t}_{AB}$$

To obtain $\hat{t}_{AB}$ we shall employ position vectors to points B and A as follows:

$$\hat{t}_{AB} = \frac{r_B - r_A}{|r_B - r_A|} = \frac{(20j + 10k) - (10i)}{\sqrt{(20)^2 + (10)^2 + (10)^2}}$$

$$= -\frac{10}{\sqrt{600}}i + \frac{20}{\sqrt{600}}j + \frac{10}{\sqrt{600}}k$$

$$= -.408i + .816j + .408k$$

We could also have determined the unit vector $\hat{t}_{AB}$ by directly ascertaining the displacement vector from A to B by inspection. Thus, to go (see Fig. 2.4) from A to B you move 20 ft in the y direction, 10 ft in the minus x direction, and then 10 ft in the z direction, thereby generating a displacement vector of $20j - 10i + 10k$. Then the unit vector $\hat{t}_{AB}$ is readily determined.

We can now give the force F as follows:

$$F = -408i + 816j + 408k \text{ lb}$$

The desired moment then becomes:

$$M_E = \rho_{EA} \times F = -20j \times (-408i + 816j + 408k)$$

$$= -8160k - 8160i \text{ lb-ft}$$

2.3. Moment of a Force about an Axis

To compute the moment of a force F about an axis BB (Fig. 2.5), we pass any plane A perpendicular to the axis so as to cut the line of action of the force at some point P. The force F is then projected to form a rectangular component F_B parallel to BB (and thus perpendicular to plane A) as shown in the diagram. The intersection of

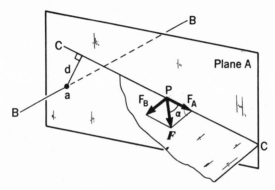

FIGURE 2.5

plane A with the plane of forces F_B and F gives a direction CC along which the other rectangular component of F denoted as F_A can be projected. The moment of F about the line BB is then defined as the moment of the component F_A about the trace point a of the axis BB—a coplanar problem discussed at the beginning of the previous section. Thus, according to the definition, the component F_B, which is parallel to the axis BB, contributes no moment about the axis, and we may say:

$$\text{moment about the axis } BB = (F_A)(d) = |F|(\cos \alpha)(d)$$

Although this quantity is associated with a particular axis that has a distinct direction, it is a scalar. The situation is the same as it is with the scalar components V_x, V_y, F_x, etc., which are associated with certain directions but are nevertheless scalars.

By means of a simple situation, we can easily show why we set forth the above definition. Suppose a disc is mounted on a shaft that is free to rotate in a set of bearings, as shown in Fig. 2.6, and a force F inclined to the plane A of the disc acts on the disc. We decompose the force into components respectively parallel to the axis of the shaft and

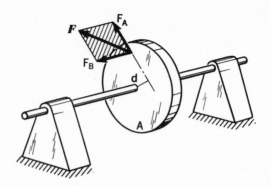

FIGURE 2.6

tangent to the plane A, that is, into forces F_B and F_A. We know from experience that F_B does not cause the disc to rotate; however, we know from physics that it is the product of F_A and the perpendicular distance d from the centerline of the shaft to the line of action of F_A that is related to the rotational motion of the system. But according to our definition, this is nothing more than the moment of force F about the axis of the shaft.

What is the relation between a moment about a point and a moment about an axis? To answer this most simply, consider BB in Fig. 2.5 to be an x axis, as shown in Fig. 2.7, and formulate the moment of a force F about this x axis in the following manner. Select any point O along the x axis as the origin of a coordinate system and decompose F into orthogonal components parallel to the reference. The component F_x normal to plane A (now a plane parallel to the yz plane) contributes no moment about the x axis. Therefore, we need to be concerned only with the components F_y and F_z. Note that the coordinates y and z of point P do not depend on the chosen position O along the x axis. Since only these coordinates will enter our discussion, the precise

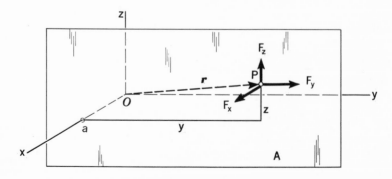

FIGURE 2.7

position of O along the x axis, i.e., the value of x, is of no interest now. Taking moments about point a, we then have, in terms of these components:

$$\text{moment about } x \text{ axis} = F_z y - F_y z$$

You will note that this result, which gives the moment of force $\boldsymbol{F}$ about the x axis in terms of its Cartesian components, is identical to the right side of Eq. 2.5(a) wherein we have the component in the x direction of the moment $\boldsymbol{M}$ of force $\boldsymbol{F}$ about the origin O of reference x, y, z. We can then conclude that:

$$\text{moment about } x \text{ axis} = M_x = \boldsymbol{M} \cdot \boldsymbol{i} \qquad \textbf{2.6}$$

We may generalize the preceding remarks to any axis n (to which we assign the unit vector $\boldsymbol{n}$) in the following manner:

$$\text{moment about } n \text{ axis} = M_n = \boldsymbol{M} \cdot \boldsymbol{n} \qquad \textbf{2.7}$$

where $\boldsymbol{M}$ is the moment of force $\boldsymbol{F}$ about any point along the axis n. *This equation stipulates* in words *that the moment of a force about an axis* equals *the scalar component in the direction of the axis of the moment vector taken about any point along the axis.*

If we specify the moments of a force about three orthogonal concurrent axes, we then single out one possible point in space for O, which is, of course, the common point of the axes. These three quantities then become the orthogonal scalar components of the moment of $\boldsymbol{F}$ about this point, and we can say:

$$
\begin{aligned}
\boldsymbol{M} = {}& (\text{moment about the } x \text{ axis})\boldsymbol{i} \\
& + (\text{moment about the } y \text{ axis})\boldsymbol{j} \qquad \textbf{2.8} \\
& + (\text{moment about the } z \text{ axis})\boldsymbol{k} = M_x \boldsymbol{i} + M_y \boldsymbol{j} + M_z \boldsymbol{k}
\end{aligned}
$$

*From this relation we can conclude that the orthogonal components of the moment of a force about a point are the moments of this force about the orthogonal axes that have the point as an origin.**

---

*You may now ask what are the physical differences in applications of moments about an axis and moments about a point. The simplest example is in the dynamics of rigid bodies. If an object is constrained so it can only spin on an axis, as in Fig. 2.6, the rotary motion will depend on the moment of the forces about the axis of rotation, as related by a scalar equation. The less familiar concept of moment about a point is illustrated in the motion of bodies such as rockets that have no constraints. In these cases, the motion of the body is related by a vector equation to the moment of forces acting on the body about the center of mass (to be defined later).

EXAMPLE 2.2

Compute the moment of a force $F = 10i + 6j$ lb, which goes through position $r_a = 2i + 6j$ ft, about a line going through points (1) and (2) having respective position vectors:

$$r_1 = 6i + 10j - 3k \text{ ft}$$
$$r_2 = -3i - 12j + 6k \text{ ft}$$

To compute this moment we can take the moment of F about either point (1) or (2) and then find the component of this vector along the direction of the displacement vector between (1) and (2). Mathematically we have, using point (1),

$$M_\rho = [(r_a - r_1) \times F] \cdot \hat{\rho} \tag{a}$$

where $\hat{\rho}$ is the unit vector along the direction between points (1) and (2). The above calculation can readily be carried out after the vectors have been expressed as scalar components. Thus we have:

$$r_a - r_1 = (2i + 6j) - (6i + 10j - 3k)$$
$$= -4i - 4j + 3k$$
$$F = 10i + 6j$$
$$\hat{\rho} = \frac{r_1 - r_2}{|r_1 - r_2|} = \frac{9i + 22j - 9k}{\sqrt{81 + 484 + 81}}$$
$$= 0.355i + 0.870j - 0.356k$$

If you recognize Eq. (a) to be a scalar triple product (see appendix), then you may solve for M_ρ by the following simple determinental approach:

$$M_\rho = \begin{vmatrix} -4 & -4 & 3 \\ 10 & 6 & 0 \\ 0.355 & 0.870 & -0.356 \end{vmatrix} = 14.02 \text{ lb-ft} \tag{b}$$

2.4. The Couple and Couple-Moment

A special arrangement of forces that is of great impor-tance is the couple. *The couple is formed by any two equal parallel forces that have opposite senses* (Fig. 2.8). On a rigid body it has only one effect,

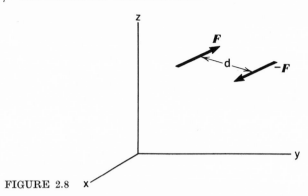

FIGURE 2.8

namely, a "twisting" action. Non-couple combinations of forces may "push" as well as "twist" a body. In either the couple or non-couple cases this twisting action is given quantitatively by the moment of forces about a point or line as mentioned in the footnote of Section 2.3. We shall accordingly be most concerned with the moment of a couple, or what we shall call the *couple-moment*.

Let us now evaluate the moment of the couple about the origin. Position vectors have been drawn in Fig. 2.9 to points A and B somewhere along the line of action of each force. Adding the moment of each force about O we have:

$$M = r_1 \times F + r_2 \times (-F)$$
$$= (r_1 - r_2) \times F \qquad\qquad 2.9$$

We can see that $(r_1 - r_2)$ is a displacement vector between points A and B, and if we call this vector e, the above formulation becomes:

$$M = e \times F \qquad\qquad 2.10$$

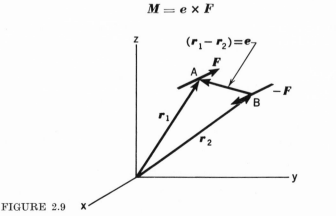

FIGURE 2.9

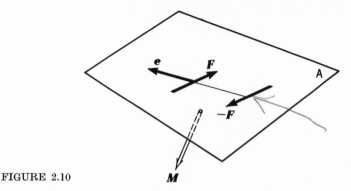

FIGURE 2.10

Since e is in the plane of the couple, it is clear from the definition of a cross product that M has a line of action normal to the plane of the couple. The sense in this case may be seen in Fig. 2.10 to be directed downward, in accordance with the right-hand screw rule. Note that the rotation of e to F, as stipulated in the cross-product formulation, is in the same direction as the "twist" of the two force vectors, and from now on we shall use the latter criterion for determining the direction of rotation to be used with the right-hand screw rule.

Now that the direction of M has been established for the couple, for a complete description we need only compute the magnitude. Points A and B may be chosen anywhere along the lines of action of the forces without changing the resulting moment, since the forces are transmissible for this computation. Therefore, to compute the magnitude of the couple vector it will be simplest to choose positions A and B so that e is perpendicular to the lines of action of the forces (e is then denoted as $e_\perp$). From the definition of the cross product, we can then say:

$$|M| = |e_\perp||F| \sin 90° = |e_\perp||F| = |F|d \qquad \textbf{2.11}$$

where the more familiar notation, d, has been used in place of $|e_\perp|$ as the perpendicular distance between the lines of action of the forces.

To summarize the preceding discussions, we may say that the moment of a couple is a vector whose line of action is normal to the plane of the couple and whose sense is determined in accordance with the right-hand screw rule, using the "twist" of the forces to give the proper rotation. The magnitude of the couple-moment equals the product of the force magnitudes comprising the couple times the perpendicular distance between the forces.

2.5. The Couple-Moment as a Free Vector

Had we chosen any other position in space as the origin O' and had we computed the moment of the couple about it, we would have formed the very same moment vector. To understand this, note that although the position vectors to points A and B will change for a new origin, the *difference* between these vectors (which has been termed e) does *not* change, as can readily be observed in Fig. 2.11. Since $M = e \times F$, we can conclude that *the couple has the same moment about every point in space*. The particular line of action of the vector representation of the couple-moment that is illustrated in Fig. 2.10 is then of little significance. In short, the *couple-moment is a free vector*. That is, we may move this vector anywhere in space without changing its meaning, provided we keep the direction and magnitude intact. Consequently, *for the purpose of taking moments* we may move the couple itself anywhere in its own or parallel plane, provided the direction of twist is not altered.* In any of these possible planes, we can also change the magnitude of the forces of the couple to other equal values, provided the distance d is simultaneously changed so that the product $|F|d$ remains the same. Since none of these steps changes the direction or magnitude of the couple-moment, all of them are permissible.

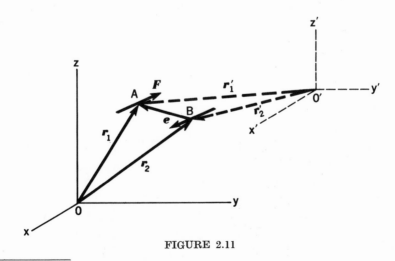

FIGURE 2.11

---

*This will not happen if we do not "flip" the couple over.

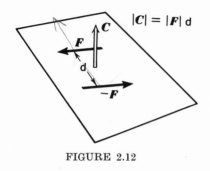

$$|C| = |F|\,d$$

FIGURE 2.12

As was pointed out earlier, the major effect of a couple is its twisting action, which is represented quantitatively by the moment of the couple. Since this is so often its sole effect, it is only natural to represent the couple by specifications of its moment; its magnitude, then, becomes $|F|\,d$ and its direction that of its moment. This is the same as identifying a man by his job, i.e., as a teacher, plumber, etc. Thus in Fig. 2.12, C is used to represent the indicated couple.

2.6. The Addition and Subtraction of Couples

The addition and subtraction of couples actually mean the addition and subtraction of the moments of the couples, for the reasons given above. Since couple-moments are free vectors, we can always arrange to form a concurrent system of vectors, which we learned to add in earlier course work in physics. We shall nevertheless take the opportunity to illustrate many of the earlier remarks about couples by adding the two couples shown on the face of the cube in Fig. 2.13. Notice that the moment representations of the couples have been drawn. Since these vectors are free, they may be moved to a

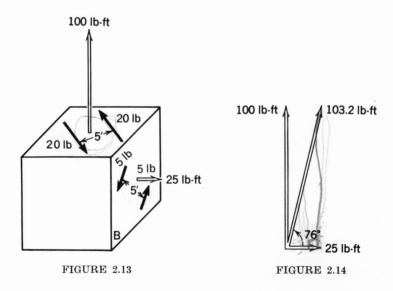

FIGURE 2.13   FIGURE 2.14

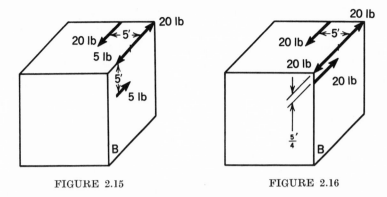

FIGURE 2.15   FIGURE 2.16

convenient position and then added. The total couple-moment then becomes 103.2 lb-ft at an angle of 76° with the horizontal as shown in Fig. 2.14. This means that the couple that creates this twisting action is in a plane at right angles to this direction with a clockwise direction as observed from below.

This addition may be shown to be valid by the following more elementary procedure. The couples of the cube are moved in their respective planes to the positions shown in Fig. 2.15. This does not alter the moment of forces, as pointed out in the previous paragraphs. If the couple on plane B is adjusted to have a force magnitude of 20 lb and if the separating distance is decreased to $\frac{5}{4}$ ft, the couple magnitude is not changed (Fig. 2.16). We thus form a system of forces in which two of the members are equal, opposite, and collinear and since these two forces cannot contribute moment, they may be deleted, leaving a single couple on a plane inclined to the original planes (Fig. 2.17). The distance between the remaining forces is:

$$\sqrt{25 + \tfrac{25}{16}} = 5.16 \text{ ft,}$$

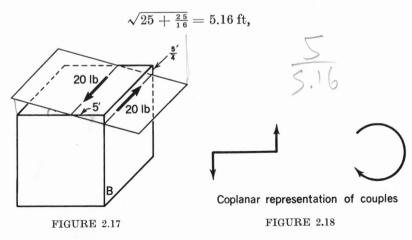

FIGURE 2.17   FIGURE 2.18

Coplanar representation of couples

and so the magnitude of the couple-moment may then be computed to be 103.2 lb-ft. The direction of the normal to the plane of the couple is readily evaluated as 76° with the horizontal, making it identical to our preceding results. Thus we see that the previous representation of the couple in terms of its moment is a convenient one for adding the couples.

In coplanar problems, a notation that is in common use to present the couple is shown in Fig. 2.18.

2.7. A Note on the Scalar Components of a Couple-Moment

In a previous section, we learned that the moment of a force about a point was a vector quantity whose scalar component along a given line of action through the point was interpreted to be the moment of the force about that line. Since the couple-moment is developed from moments of forces, the same conclusion may be reached for the scalar components of this vector. To illustrate, a couple-moment

FIGURE 2.19

C is shown in Fig. 2.19 as well as an arbitrary direction s. We may then say that $C_s = C \cdot s$ where C_s is a scalar representing the moment of the couple about the line s. Since C is a free vector, the moments about all parallel lines of a given set are equal.

2.8. Equivalent Force Systems

In Chapter 1 we defined equivalent vectors as those that have the same capacity in some given situation. We will now investigate an important class of situations, namely those in which a *rigid-body model* can be employed, and we will be concerned with the question of what are the equivalence requirements for force systems acting on a rigid body.

The effect that forces have on a rigid body is only manifested in the motion (or lack of motion) of the body induced by the forces. Two force systems, then, are equivalent if they are capable of initiating the same motion in the rigid body. The necessary and sufficient conditions required to give the force systems this equal capacity are:

  a. Each force system must exert an equal "push" or "pull" on the body in any direction. For concurrent force systems, this requirement is satisfied if the addition of the force systems results in equal vectors.
  b. Each force system must exert an equal "twist" about any point in space. This means that the moment vectors of the force systems for any chosen point must be equal.

Although these conditions will most likely be intuitively acceptable, you can later prove their validity when you study dynamics, starting with Newton's second law.

We shall reiterate several basic force equivalences for rigid bodies that will serve as a foundation for more complex considerations. You should subject them to the tests listed above.

  1. The sum of a set of concurrent forces is a single force that is equivalent to the original system. Conversely, a single force is equivalent to any set of its components that is concurrent to it.
  2. A force may be moved along its line of action, i.e., forces are transmissible vectors.
  3. Since a couple is a free vector, for our present purposes the couple may be altered in any way as long as the couple-moment is not changed.

In succeeding sections we shall present other equivalence relations for rigid bodies and then examine perfectly general force systems with a view to replacing them with more convenient and simpler equivalent force systems. These simpler replacements are often called *resultants* of the more general systems.

2.9. Translation of a Force to a Parallel Position

In Fig. 2.20 let us consider the possibility of moving a force F (full line) to a parallel position at point a while maintaining rigid-body equivalence. If we add equal and opposite forces F and $-F$ at position a, a system of three forces is formed that is clearly equivalent to the original single force F since the additional equal and

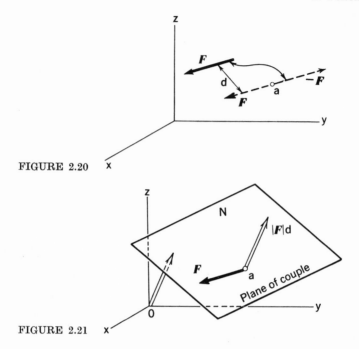

FIGURE 2.20

FIGURE 2.21

opposite forces cancel each other's effects. Note that the original force *F* and an added force in the opposite sense form a couple (the pair is identified by a wavy connecting line). As usual we represent the couple by its moment, as is shown in Fig. 2.21, where the plane of the newly formed couple is noted as *N*. The magnitude of the couple is $|F|d$ where *d* is the perpendicular distance between point *a* and the original line of action of the force. The couple-moment may then be moved to any parallel position, including the origin, as is indicated in Fig. 2.21.

Thus we see that a force may be moved to any parallel position, provided a couple of the correct orientation and size is simultaneously developed. There are, then, an infinite number of arrangements possible to get the equivalent effects of a force on a rigid body.

The reverse procedure may also be instituted in reducing a force and a couple in the same plane to a single equivalent force. This is

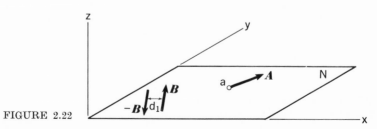

FIGURE 2.22

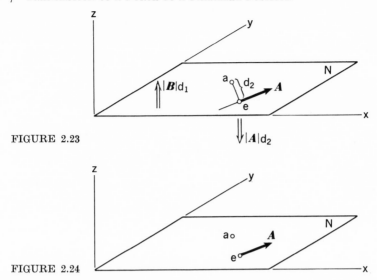

FIGURE 2.23

FIGURE 2.24

illustrated in Fig. 2.22 where a couple composed of forces B and $-B$ and a force A are shown in plane N. In Fig. 2.23 we have used the couple-moment representation of the forces B and we have shifted force A a distance d_2 in plane N to go through point e such that a new couple-moment $|A|d_2$ is formed that is equal and opposite to the original one $|B|d_1$. The couple-moments then cancel, leaving a single force A in plane N going through point e, as has been shown in Fig. 2.24.*

A simple approach for establishing the proper couple-moment when moving a force is to make use of the cross product. You will recall that the moment of a couple about any point fully establishes the couple-moment. Hence, returning to Fig. 2.20, we determine the couple-moment of the newly formed couple by taking the moment of the couple about point a. But this moment is nothing more than the moment of the original force F about point a. *Thus in shifting a force to pass through some new point we introduce a couple whose couple-moment equals the moment of the force about this point.* This is shown in Fig. 2.25 where in part (a) we have shown a force F and point a forming plane A. A displacement vector ρ is drawn from a to any point along the line of action of F. In part (b) of the diagram we have shown the equivalent force system with F going through a. The couple moment is $\rho \times F$ and clearly must be normal to plane A.

---

*If force A were not coplanar with the plane of the couple and therefore had a component in the z direction A_z, then the reduction to a single force would not be possible. The reason for this is that in going from a to e to eliminate $|B|d_1$ as above, the component A_z would generate a couple-moment *parallel* to the xy plane.

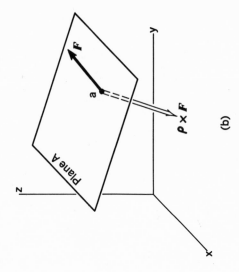

(a)

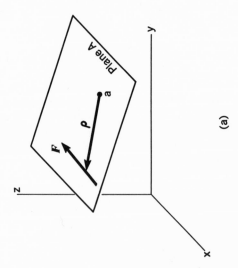

(b)

FIGURE 2.25

EXAMPLE 2.3

A force $F = 6i + 3j + 6k$ lb goes through a point whose position vector is $r_1 = 2i + j + 10k$ ft. Replace this force by an equivalent force system for purposes of rigid-body mechanics going through position $r_2 = 6i + 10j + 12k$ ft.

The new system will consist of the force F going through the position r_2 and, in addition, there will be a couple-moment given as:

$$M = \rho \times F = (r_1 - r_2) \times F$$

Inserting values we have:

$$M = [(2i + j + 10k) - (6i + 10j + 12k)] \times (6i + 3j + 6k)$$
$$= (-4i - 9j - 2k) \times (6i + 3j + 6k)$$
$$= -48i + 12j + 42k \text{ lb-ft}$$

2.10. Resultant of a Force System

As defined earlier, a *resultant of a force system* is a simpler equivalent force system. You will find in many problems that it is desirable first to establish the resultant before entering into other computations.

For a general arrangement of forces, no matter how complex, we may always move all forces to proceed through a single point. The original couple-moments and the couple-moments induced by moving forces may then also be placed to go through the point. The result is a system of concurrent forces and a system of concurrent couple-moments at the point. These may then be combined into a single force and a single couple. Thus in Fig. 2.26 we have shown some arbitrary system of forces and couples using full lines. The resultant force and

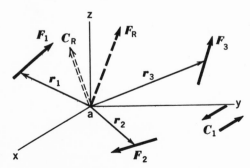

FIGURE 2.26

couple-moment combination is shown as dashed lines at the origin of a rectangular reference drawn at point a.

The methods of finding a resultant of forces involve nothing new. In moving to *any* new point, you will recall, there is no change in the force itself other than a shift of line of action; thus any component of the *resultant* force, such as the x component, can simply be taken as the sum of the respective x components of all the forces in the system. We may then say for the resultant force:

$$\boldsymbol{F}_R = [\sum_i (F_x)_i]\boldsymbol{i} + (\sum_i (F_y)_i]\boldsymbol{j} + [\sum_i (F_z)_i]\boldsymbol{k} \qquad \textbf{2.12}$$

The couple-moment accompanying $\boldsymbol{F}_R$ for a chosen point a may then be given as:

$$\boldsymbol{C}_R = [\boldsymbol{\rho}_1 \times \boldsymbol{F}_1 + \boldsymbol{\rho}_2 \times \boldsymbol{F}_2 + \cdots + \boldsymbol{\rho}_n \times \boldsymbol{F}_n] + [\boldsymbol{C}_1 + \cdots + \boldsymbol{C}_m] \quad \textbf{2.13}$$

where the first bracketed quantities result from moving the noncouple forces to a and the second are simply the sum of the given couples. The vectors $\boldsymbol{\rho}$ are from a to arbitrary points along the line of action of the forces. In more compact form the above equation becomes:

$$\boldsymbol{C}_R = \sum_{i=1}^{n} \boldsymbol{\rho}_i \times \boldsymbol{F}_i + \sum_{i=1}^{m} \boldsymbol{C}_i \qquad \textbf{2.14}$$

The following example illustrates the procedure.

EXAMPLE 2.4

Shown in Fig. 2.27 are two forces and a couple, the couple being positioned in plane zy. We shall find the resultant of the system at the origin O.

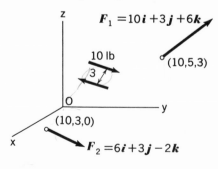

$F_1 = 10\boldsymbol{i} + 3\boldsymbol{j} + 6\boldsymbol{k}$

10 lb

(10,5,3)

(10,3,0)

$F_2 = 6\boldsymbol{i} + 3\boldsymbol{j} - 2\boldsymbol{k}$

FIGURE 2.27

At O we will have a set of two concurrent forces, which may be added to give $\boldsymbol{F}_R$:

$$\boldsymbol{F}_R = (10 + 6)\boldsymbol{i} + (3 + 3)\boldsymbol{j} + (6 - 2)\mathbf{k}$$
$$= 16\boldsymbol{i} + 6\boldsymbol{j} + 4\boldsymbol{k}$$

The couple-moment at this point is the vector sum of the moment vectors developed by moving the two forces, plus the 30 ft-lb couple in the zy plane. Thus:

$$C_R = r_1 \times F_1 + r_2 \times F_2 - 30i$$

Now:

$$r_1 \times F_1 = (10i + 5j + 3k) \times (10i + 3j + 6k)$$
$$= 21i - 30j - 20k$$
$$r_2 \times F_2 = (10i + 3j) \times (6i + 3j - 2k)$$
$$= -6i + 20j + 12k$$

Hence:

$$C_R = -15i - 10j - 8k$$

The resultant is shown in Fig. 2.28.

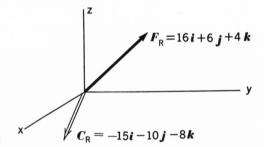

FIGURE 2.28

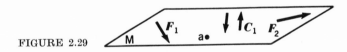

FIGURE 2.29

We often encounter forces and couples that are *coplanar* (see Fig. 2.29). We can readily find a resultant force system at any point a in the plane by summing forces (Eq. 2.12) as explained above. The accompanying couple for point a (Eq. 2.14) will be normal to the plane of the forces. If the sum of the forces is *not zero*, we can simplify the system further by moving the resultant force to a new position a' so as to induce further a couple that will cancel the first couple—a process that has been explained earlier for a force and a couple-moment at right angles to each other. The simplest resultant for this case is then a single force passing through the position a'. On the other hand, if the forces sum to a zero vector, then clearly the resultant must be a couple (or be zero). In the following examples we shall compute simplest resultants.

EXAMPLE 2.5

Consider a coplanar force system shown in Fig. 2.30. The *simplest* resultant is to be found. Since $\sum F_x$ and $\sum F_y$ are not zero, we know that we can replace the system by a single force, which is:

$$F_R = 6i + 13j$$

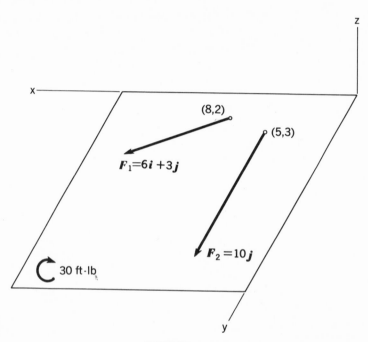

FIGURE 2.30

We next find the line of action of F_R with no couple. To do this we move the forces to go through point a', whose position we denote as $\bar{x}i + \bar{y}j$ (see Fig. 2.31). We now set $C_R = 0$ in Eq. 2.14 for this point measuring ρ_i from the point to the forces as shown in the diagram. Thus we have:

$$0 = \rho_1 \times F_1 + \rho_2 \times F_2 - 30k$$
$$= [(8i + 2j) - (\bar{x}i + \bar{y}j)] \times (6i + 3j)$$
$$+ [(5i + 3j) - (\bar{x}i + \bar{y}j)] \times (10j) - 30k$$

We may obtain the following equation:

$$13\bar{x} - 6\bar{y} = 32$$

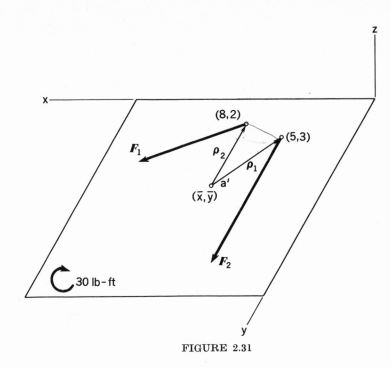

FIGURE 2.31

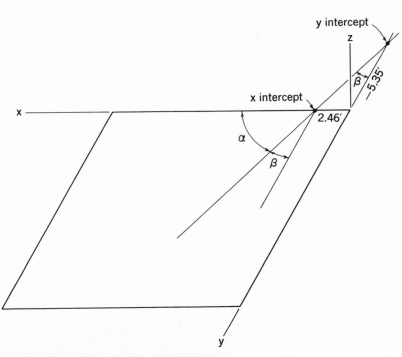

FIGURE 2.32

41

We have here a locus of points for a' along a straight line. This line clearly must then represent the line of action of the simplest resultant force for this problem. (We leave it to you to show, using Fig. 2.32 as an aid, that the straight line given above has identically the same orientation as the resultant force.) Usually we merely specify one of the intercepts for such a case. To do this we place a' on the x axis so that the position vector r' of a' is simply $\bar{x}i$. We can then see from the above computations that:

$$\bar{x} = 2.46 \text{ ft}$$

The result is shown in Fig. 2.33.

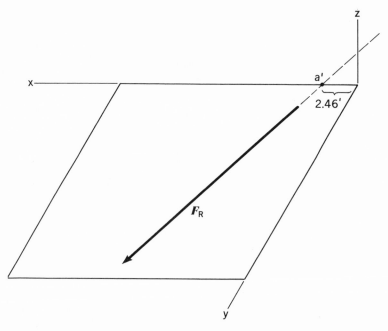

FIGURE 2.33

A better way to proceed is to use the fact that the resultant is *equivalent* to the system as far as moments are concerned. We then equate the moment about the origin (or some other convenient point) of the simplest resultant (no couple) with that of the system. Thus, using the yet-to-be-determined intercept of the resultant with the x axis for determining the moment arm for the resultant (see Fig. 2.34), we obtain:

$$\bar{x}i \times (6i + 13j) = (8i + 2j) \times (6i + 3j) + (5i + 3j) \times 10j - 30k$$

The obtained value of $\bar{x}$ is again 2.46, as you may yourself verify.

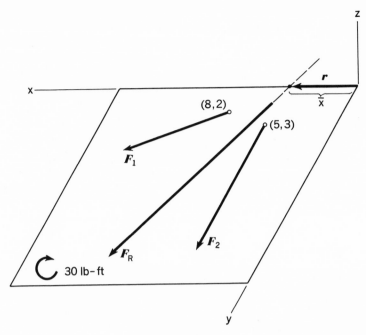

FIGURE 2.34

We have thus presented two approaches. The first is more formal, following the basic formulations of resultants, while the second makes direct use of equivalance and perhaps affords greater physical motivation.

EXAMPLE 2.6

In Fig. 2.35 we have shown a system of coplanar forces and a couple. The forces and a couple are given as follows:

$$F_1 = -4i + 3j \, \text{lb}$$
$$F_2 = -2i - 2j \, \text{lb}$$
$$F_3 = 6i - j \, \text{lb}$$
$$C = 20k \, \text{lb-ft}$$

We desire the simplest resultant. In summing the forces we see that $F_R = 0$ and so the resultant must be either a couple or zero. We may employ Eq. 2.14 to obtain C_R. Thus:

$$C_R = (4i + 2j) \times (-4i + 3j) + (3i + 2j) \times (-2i - 2j)$$
$$+ (i + 4j) \times (6i - j) + 20k = 23k$$

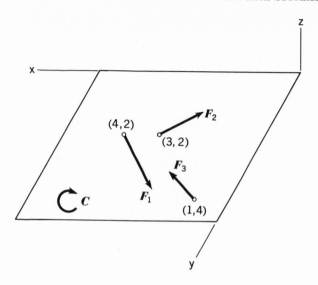

FIGURE 2.35

Another force system that appears often in engineering problems is the *parallel* force system. Such a system has been shown in Fig. 2.36. Note that we include couples whose planes are parallel to the direction of the forces (the z direction here) because such couples can be considered to be composed of equal and opposite forces parallel to the z direction.

Again, we shall be interested in obtaining the *simplest* resultant. We find F_R in Eq. 2.12 by simply adding the forces. If this sum is zero,

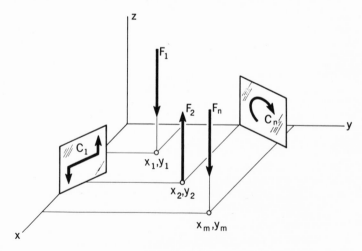

FIGURE 2.36

then the resultant will be either a couple as given by Eq. 2.14 or zero. If it is a couple, there is no way to simplify further and we have found the simplest system. If $F_R \neq 0$, then for the simplest resultant we must find the position a' in the xy plane through which F_R passes for which $C_R = 0$ in Eq. 2.14. Denoting the position vector of a' in the plane as $\bar{x}i + \bar{y}j$, we can then readily determine the coordinate $(\bar{x}, \bar{y})$ through which the resultant force with no couple passes.

We can conclude here that, just as in the case of the coplanar system, the simplest resultant force is a single force if $F_R \neq 0$ or is a couple.

EXAMPLE 2.7

Find the simplest resultant force system for the parallel forces shown in Fig. 2.37.

The resultant force is $-30k$ and so we must now determine point a' for the simplest system (see Fig. 2.38). Thus, setting $C_R = 0$ in Eq. 2.14 we obtain:

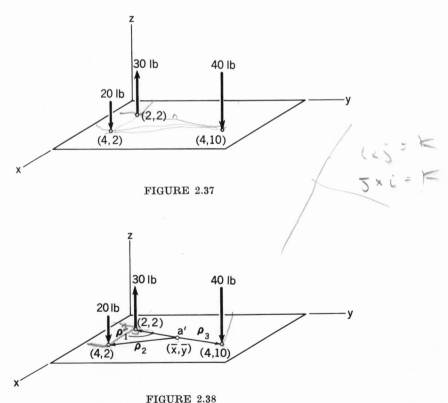

FIGURE 2.37

FIGURE 2.38

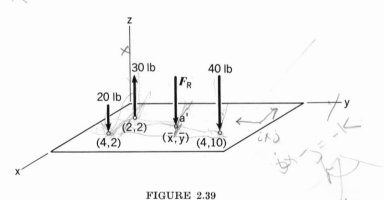# Line

46 2. FORCE SYSTEMS

$$0 = \rho_1 \times (-20k) + \rho_2 \times (30k) + \rho_3 \times (-40k)$$
$$= [(4i + 2j) - (\bar{x}i + \bar{y}j)] \times (-20k)$$
$$+ [(2i + 2j) - (\bar{x}i + \bar{y}j)] \times (30k)$$
$$+ [(4i + 10j) - (\bar{x}i + \bar{y}j)] \times (-40k)$$

This leads to the following scalar equations:

$$180 - 30\bar{x} = 0$$
$$-380 + 30\bar{y} = 0$$

Hence:

$$\bar{x} = 6 \text{ ft}; \qquad \bar{y} = 12.67 \text{ ft}$$

A simpler approach is to note that the moments about the x and y axes, respectively, of the simplest resultant (acting at $\bar{x}\bar{y}$ without a couple) equal the corresponding moments of the given force system (see Fig. 2.39). Thus, for the x axis we obtain:

$$-30\bar{y} = (2)(30) - (2)(20) - (10)(40)$$
$$\therefore \bar{y} = 12.67 \text{ ft}$$

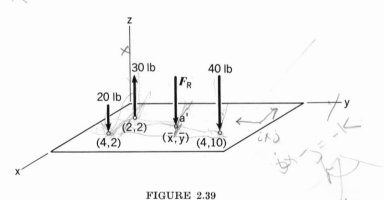

FIGURE 2.39

And for the y axis we obtain:

$$30\bar{x} = -(2)(30) + (4)(20) + (4)(40)$$
$$\therefore \bar{x} = 6 \text{ ft}$$

EXAMPLE 2.8

Consider the parallel force system in Fig. 2.40. What is the simplest resultant?

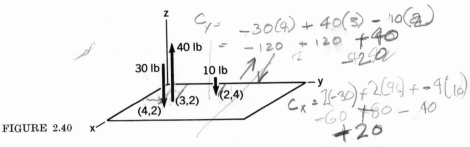

FIGURE 2.40

Here we have a case where the sum of the forces equals zero. The resultant, if not zero, must be a couple. To obtain the scalar components of the couple-moment, we take moments about the x and y axes. Thus:

$$C_y = (10)(2) - (40)(3) + (30)(4) = 20 \text{ lb-ft } \;\uparrow\!\!\jmath$$
$$C_x = -(10)(4) + (40)(2) - (30)(2) = -20 \text{ lb-ft } \;\curvearrowleft$$

The couple-moment, then, is:

$$C_R = -20\boldsymbol{i} + 20\boldsymbol{j} \text{ lb-ft}$$

We can reach the above result directly by taking moments about a convenient point such as the origin. Thus:

$$\begin{aligned} C_R &= (4\boldsymbol{i} + 2\boldsymbol{j}) \times (-30\boldsymbol{k}) + (3\boldsymbol{i} + 2\boldsymbol{j}) \times (40\boldsymbol{k}) \\ &\quad + (2\boldsymbol{i} + 4\boldsymbol{j}) \times (-10\boldsymbol{k}) \\ &= -20\boldsymbol{i} + 20\boldsymbol{j} \text{ lb-ft} \end{aligned}$$

In Chapter 3 we shall be considering the equations of equilibrium for rigid bodies. In ascertaining the number of independent equations of equilibrium for various kinds of force systems, it is helpful to keep in mind what the simplest resultant is for the particular system under consideration.

2.11. Distributed Force Systems

Our discussions up to now have been restricted to discrete vectors, in particular to the point force. Scalars and vectors may also be continuously distributed throughout a region so that at each position in the region there may be given a definite scalar or vector quantity. Such distributions are called *scalar fields* and *vector fields*, respectively. A simple example of a scalar field is the temperature distribution, expressed as $T(x, y, z, t)$ where the variable t indicates that the field may be changing with time. Thus if a position x_0, y_0, z_0 and a time t_0 are specified, we can determine the temperature at this position and time if we know the temperature distribution function. A vector field such as the force field is sometimes expressed in the form $F(x, y, z, t)$. However, in place of the vector field, it is more convenient at times to employ three scalar fields that represent the orthogonal scalar components of a vector field at all points. Thus for a force field we can say:

$$\text{force component in } x \text{ direction} = g(x, y, z, t)$$
$$\text{force component in } y \text{ direction} = h(x, y, z, t)$$
$$\text{force component in } z \text{ direction} = k(x, y, z, t)$$

where g, h, and k represent functions of the coordinates and time. If we substitute coordinates of a special position and the time into these functions, we get the force components F_x, F_y and F_z for that position and time. The force field and its component scalar fields are then related in this way:

$$F(x, y, z, t) = g(x, y, z, t)\boldsymbol{i} + h(x, y, z, t)\boldsymbol{j} + k(x, y, z, t)\boldsymbol{k}$$

More often the notation for the above equation is written as:

$$F(x, y, z, t) = F_x(x, y, z, t)\boldsymbol{i} + F_y(x, y, z, t)\boldsymbol{j} + F_z(x, y, z, t)\boldsymbol{k} \quad \textbf{2.15}$$

An example of a force field is the gravitational field about the earth, and also the electrostatic field about electric charges. Vector fields are not restricted to forces but include such other quantities as velocity fields, heat flow fields, etc.

Force distributions that exert influence directly on the elements of mass distributed throughout the body are termed *body force distributions*, and are usually given in terms of per unit of mass of the matter that they directly influence. Thus if $B(x, y, z, t)$ is such a body force distribution, the force on an element dm would be $B(x, y, z, t) \, dm$.

Force distributions over a *surface* are called *surface force distributions* and are given in terms of per unit area of the surface directly influenced. A simple example is the force distribution on the surface of a body submerged in a fluid. In the case of a static fluid or of a frictionless fluid, the force from the fluid on an area element is always normal to the area element and directed in toward the body. The force per unit area stemming from such fluid action is called *pressure* and is denoted as p. Like force components, pressure is a scalar quantity. The direction of the force resulting from a pressure on a surface is given by the orientation of the surface. (You will recall from earlier studies that an area element can be considered as a vector which is normal to the area element and which is directed outward from the enclosed body (Fig. 2.41).) The infinitesimal force on the area element is then given as:

$$df = -p\, dA$$

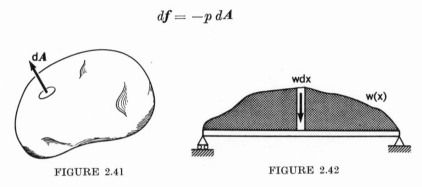

FIGURE 2.41   FIGURE 2.42

Still a more specialized, but nevertheless common, surface force distribution is that of a continuous load on a beam. This is often a parallel loading distribution that is symmetrical about the center plane of a beam, as illustrated in Fig. 2.42. We can therefore replace it by an equivalent coplanar distribution that acts at the center plane. The loading is given in terms of per unit length of the beam, and is denoted as w, the *intensity of loading*. The force on an element of the beam, then, is wdx.

We have thus presented force systems distributed throughout volumes, over surfaces, and over lines. The conclusions about resultants that were reached earlier for general, parallel, and coplanar point-force systems are also valid for these distributed-force systems. This statement is true because each system can be considered as an infinite number of infinitesimal point forces of the type developed in the previous paragraphs. We shall illustrate this fact in the following examples.

CASE A. PARALLEL BODY FORCE SYSTEM—CENTER OF GRAVITY. Consider a rigid body (Fig. 2.43) whose density (mass/unit volume) is given as $\rho(x, y, z)$. It is acted on by gravity, which may be considered to result in a distributed parallel force system.

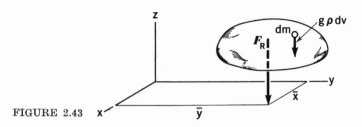

FIGURE 2.43

Since we have here a parallel system of forces in space with the same sense, we know that a single force along a certain line of action will be equivalent to the distribution. The body force $B(x, y, z)$ given per unit mass is $-g\boldsymbol{k}$. The infinitesimal force on a differential mass element dm, then, is $-g(\rho\, dv)\boldsymbol{k}$ where dv is the volume of the element. We find the resultant force on the system as follows:

$$\boldsymbol{F}_R = -\int_V g(\rho\, dv)\, \boldsymbol{k} = -g\boldsymbol{k} \int_V \rho\, dv = -gM\boldsymbol{k}$$

where with g as a constant the second integral becomes simply the entire mass of the body M.

Next we must find the line of action of this single equivalent force. Let us denote the intercept of the line of action of this force with the xy plane as $\bar{x}, \bar{y}$. The resultant at this position must have the same moments as the distribution about both the x and y axes. We can solve for these coordinates in the following manner:

$$F_R\bar{x} = -g\int_V x\rho\, dv, \qquad F_R\bar{y} = -g\int_V y\rho\, dv$$

Hence we have:

$$\bar{x} = \frac{\int x\rho\, dv}{M}, \qquad \bar{y} = \frac{\int y\rho\, dv}{M}$$

Thus we have fully established the resultant. Now the body is reoriented in space, keeping with it the line of action of the resultant as shown in Fig. 2.44. A new computation of the line of action of a resultant for the second orientation yields one that intersects the original line at a point C. It can be shown the lines of action for all other orientations of the body must intersect at the same point. We call this point the *center of gravity*. Effectively we can say for rigid-body considerations that all the weight of the body can be assumed concentrated at the center of gravity.

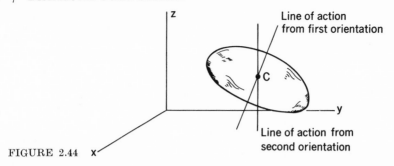

FIGURE 2.44

EXAMPLE 2.9

Find the center of gravity of the triangular block shown in Fig. 2.45 having a uniform density ρ.

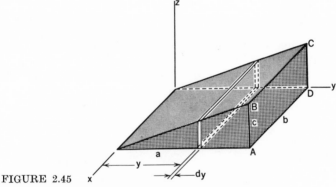

FIGURE 2.45

The total weight of the body is easily evaluated as:

$$F_R = g\rho\,\frac{abc}{2} \tag{a}$$

To find $\bar{y}$ we employ infinitesimal slices of thickness dy parallel to the xz plane as shown in the diagram. Equating moments of the weights of these slices about the x axis with the moment of the total weight of the block about the x axis, we get:

$$\bar{y}\left(g\rho\,\frac{abc}{2}\right) = g \int_0^a \rho y(zb\,dy) \tag{b}$$

The term z can be expressed as follows:

$$z = \left(\frac{y}{a}\right)c \tag{c}$$

We then have for Eq. (b):

$$\bar{y} = \frac{1}{g\rho\frac{abc}{2}}g\int_0^a \rho y^2 \frac{bc}{a}\,dy = \frac{2}{3}a \qquad (d)$$

To find $\bar{z}$, we could reorient the body so that $ABCD$ becomes the bottom face. A computation similar to the preceding one would give the result that $\bar{z} = \frac{2}{3}c$. You are urged to verify this yourself.

Finally, it should be clear by inspection that $\bar{x} = \frac{1}{2}b$.

EXAMPLE 2.10

Find the center of gravity for the body of revolution shown in Fig. 2.46. The radial distance of the surface from the y axis is given as:

$$r = \tfrac{1}{20}y^2 \qquad (a)$$

The body is 10 ft long and has cylindrical hole at the right end of length 2 ft and diameter 1 ft.

FIGURE 2.46

We need only compute $\bar{y}$, since it is clear that $\bar{z} = \bar{x} = 0$ owing to symmetry. We first compute the weight of the body. Using a slice of thickness dy as shown in the diagram we proceed as follows:

$$W = g\rho \left\{ \int_0^{10} \pi r^2\,dy - \frac{\pi(1)^2}{4}(2) \right\} \qquad (b)$$

Using Eq. (a) to replace r^2 we get:

$$W = \rho g \left[\pi \int_0^{10} \frac{y^4}{400}\,dy - \frac{\pi}{2} \right] = g\rho\pi\left(50 - \frac{1}{2}\right)$$

$$W = 49.5\pi\rho g \text{ lb} \qquad (c)$$

To get $\bar{y}$ we now proceed by equating moments about the x axis as follows:

$$(49.5\pi\rho g)\bar{y} = g\rho\left\{\int_0^{10} y\pi r^2\, dy - (9)\left(\frac{\pi}{2}\right)\right\}$$

$$\bar{y} = \frac{1}{49.5}\left\{\int_0^{10} \frac{y^5}{400}\, dy - 4.5\right\} = 8.31 \text{ ft}$$

CASE B. PARALLEL FORCE DISTRIBUTION OVER A SURFACE—
CENTER OF PRESSURE. Let us now consider a pressure distribution over
a plane surface A in the xy plane in Fig. 2.47. The vertical ordinate is

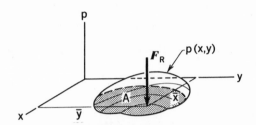

FIGURE 2.47

taken as a pressure ordinate, so that over the area A we have a pressure
distribution $p(x, y)$ represented by the pressure surface. Since in this
case there is a parallel force system, we know that the resultant is a
single force, which is given as:

$$F_R = -\int p\, dA = -\left(\int p\, dA\right)k \qquad\qquad \textbf{2.16}$$

The position $\bar{x}$, $\bar{y}$ can be computed by equating the moments of the
resultant force about the x and y axes with the corresponding moments
of the distribution:

$$\bar{x} = \frac{\int px\, dA}{\int p\, dA}$$

$$\bar{y} = \frac{\int py\, dA.}{\int p\, dA}$$

Since we know that p is a function of x and y over the surface, we can
carry out the above integrations either analytically or numerically.
The position thus established in the absence of point forces is called the
center of pressure.

In the problems at the end of this chapter we shall consider curved
surfaces with normal pressure distributions. Also, we shall consider the

distributed frictional effects over surfaces in Chapter 3. In these cases, the resultant is not necessarily a single force as it was in the above simplified case that has served primarily as an introduction.

EXAMPLE 2.11

Shown in Fig. 2.48 is a plate $ABCD$ on which there are both distributed and point forces forming a parallel system. The pressure distribution is given as:

$$p = -4y^2 + 100 \text{ psf} \tag{a}$$

and the 500-lb force acts at position $(2, 2)$. Find the simplest resultant for the system.

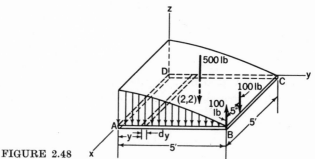

FIGURE 2.48

Summing forces to get F_R we have:

$$
\begin{aligned}
F_R &= -\int_0^5 p(5dy) - 500 \\
&= -\int_0^5 (-4y^2 + 100)5dy - 500 \\
&= \left(20\,\frac{y^3}{3} - 500y \right)\Big|_0^5 - 500 = -2170 \text{ lb} \tag{b}
\end{aligned}
$$

The simplest resultant is then a single force. Taking moments about the x axis we have for $\bar{y}$ for this force:

$$
\begin{aligned}
-2170\bar{y} &= -\int_0^5 yp(5dy) - (500)(2) \\
&= -\int_0^5 5y(-4y^2 + 100)\,dy - 1000 \\
&= \left(20\,\frac{y^4}{4} - 500\,\frac{y^2}{2} \right)\Big|_0^5 - 1000 = -4\overset{.}{1}25 \\
\therefore \bar{y} &= 1.90 \text{ ft} \tag{c}
\end{aligned}
$$

Now taking moments about the y axis we get for $\bar{x}$:

$$2170\bar{x} = \int_0^5 \frac{5}{2}\, p(5dy) + (500)(2) - 500$$

$$= \frac{25}{2} \int_0^5 (-4y^2 + 100)dy + 500 = 4667$$

$$\therefore \bar{x} = 2.15 \text{ ft} \tag{d}$$

CASE C. COPLANAR PARALLEL FORCE DISTRIBUTION. As we pointed out earlier, this type of loading may be considered for beam problems involving material that is loaded symmetrically over the longitudinal midplane of the beam; it is represented by a loading curve $w(x)$ as shown in Fig. 2.42. This coplanar parallel force distribution may be replaced by a single force given as:

$$\boldsymbol{F}_R = -\int w(x)\, dx\, \boldsymbol{k}$$

We find the position of $\boldsymbol{F}_R$ with no couple by equating the moment of the distribution about a convenient point of the beam, usually one of the ends, with that of $\boldsymbol{F}_R$. Thus:

$$\bar{x} = \frac{\int xw(x)\, dx}{\int w(x)\, dx}$$

EXAMPLE 2.12

A simply supported beam is shown in Fig. 2.49 supporting a 1000-lb point force, a 500 lb-ft point couple, and a coplanar distributed load w lb/ft which is parabolic. Find the simplest resultant of this force system.

To express the intensity of loading in the coordinate system shown in the diagram we begin with the general formulation:

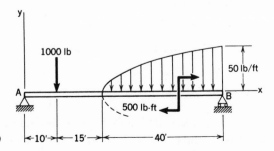

FIGURE 2.49

$$w^2 = ax + b \qquad \textbf{(a)}$$

Note that when $x = 25$ we have $w = 0$ and when $x = 65$ we have $w = 50$. Subjecting Eq. (a) to these conditions we can determine a and b. Thus:

$$0 = a(25) + b \qquad \textbf{(b)}$$

$$2500 = a(65) + b \qquad \textbf{(c)}$$

Subtracting, we can get a as follows:

$$-2500 = -40a$$

$$\therefore a = 62.5$$

From Eq. (b) we get b as:

$$b = -(25)(62.5) = -1560$$

Thus we have:

$$w^2 = 62.5x - 1560 \qquad \textbf{(d)}$$

Summing forces, we get for F_R:

$$F_R = -1000 - \int_{25}^{65} \sqrt{62.5x - 1560}\ dx$$

$$= -1000 - \frac{2}{3}\frac{1}{62.5}(62.5x - 1560)^{3/2}\Big|_{25}^{65}$$

$$= -2332\ \text{lb}$$

We now compute $\bar{x}$ for the simplest resultant as follows:

$$-2332\bar{x} = -(10)(1000)$$

$$- \int_{25}^{65} x\sqrt{62.5x - 1560}\ dx - 500$$

$$= -10,000$$

$$+ \frac{2(-3120 - 187.5x)(62.5x - 1560)^{3/2}}{(15)(62.5)^2}\Big|_{25}^{65} - 500$$

$$= -75,900$$

$$\therefore \bar{x} = 32.5\ \text{ft}$$

2.12. *Closure*

We now have the tools that enable us to replace, for purposes of rigid-body mechanics, any system of forces by a resultant

consisting of a force and a couple, and these tools will prove very helpful
in our computations. More important at this time, however, is the fact
that in considering conditions of equilibrium for rigid bodies we need
only concern ourselves with this simple system to reach conclusions
valid for any force system, no matter how complex. From this viewpoint, we shall develop the fundamental equations of statics in the
following chapter, and employ them for solving a large variety of
problems.

Problems

1. [2.1] What is the position vector r from the origin $(0, 0, 0)$ to the point $(3, 4, 5)$? What are its magnitude and direction cosines?

2. [2.1] What is the displacement vector from position $(6, 13, 7)$ to position $(10, -3, 4)$?

3. [2.1] Shown in Fig. 2.50 are two sets of parallel rectangular references XYZ and xyz. The origin of xyz is displaced from XYZ by the position vector

$$R = 10i + 6j + 5k$$

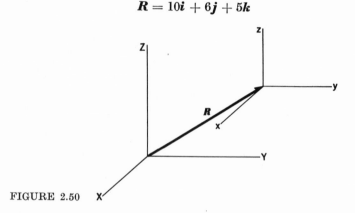

FIGURE 2.50

The position vector r' of a point relative to the XYZ reference is given as:

$$r' = 3i + 2j - 6k$$

What is the position vector r of this point for the reference xyz?

4. [2.1] A point moves along a straight line from position

$$r_1 = 10i + 6j - 3k \text{ ft}$$

to position

$$r_2 = 3i - 4j + 6k \text{ ft}$$

What is the distance traveled between these points and the direction traveled in terms of direction cosines?

58

5. [2.1] A particle moves along a circular path in the xy plane as shown in Fig. 2.51. What is the position vector $\mathbf{r}$ of this particle as a function of the coordinate x?

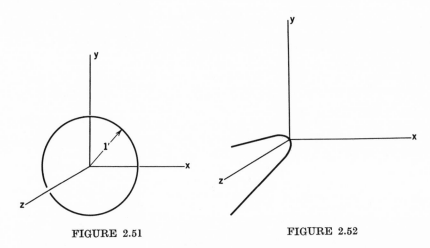

FIGURE 2.51 FIGURE 2.52

6. [2.1] A particle moves along a parabolic path in the yz plane as shown in Fig. 2.52. If the particle has at one point a position vector $\mathbf{r} = 4\mathbf{j} + 2\mathbf{k}$ give the position vector at any point on the path as a function of the z coordinate.

7. [2.1] Reference xyz is rotated 30° about its x axis as shown in Fig. 2.53 relative to reference XYZ. What is the position vector $\mathbf{r}$ for reference xyz of a point having a position vector $\mathbf{r}'$ for reference XYZ given as:

$$\mathbf{r}' = 6\mathbf{i}' + 10\mathbf{j}' + 3\mathbf{k}'$$

Use $\mathbf{i}, \mathbf{j}, \mathbf{k}$ (no primes) for unit vectors associated with reference xyz.

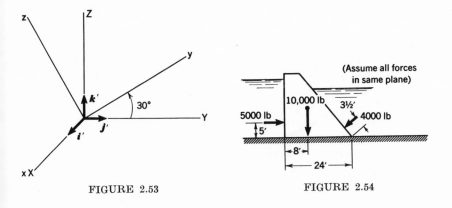

FIGURE 2.53 FIGURE 2.54

▶ In the simpler coplanar problems that follow it is suggested that you proceed using the elementary definition as well as the cross-product definition of the moment.

8. [2.2] The total equivalent forces from water and gravity are shown on the dam (Fig. 2.54). (We shall soon be able to compute such equivalents.) Compute the moment of these forces about the toe of the dam at the right-hand corner.

9. [2.2] Find the moment of the forces shown about the centerline of the step pulley at O in Fig. 2.55.

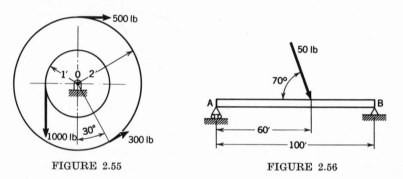

FIGURE 2.55   FIGURE 2.56

10. [2.2] Find the moment of the 50-lb force about the support at A (Fig. 2.56).

11. [2.2] Compute the moment of the 1000-lb force in Fig. 2.57 about points A, B, and C. Use the transmissibility property of force here and rectangular components to make the computations simplest.

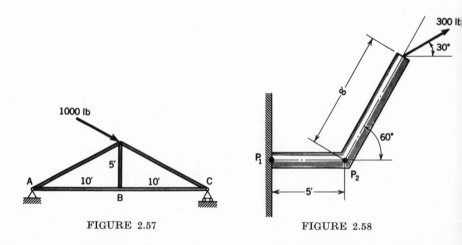

FIGURE 2.57   FIGURE 2.58

12. [2.2] A force $\boldsymbol{F} = 10\boldsymbol{i} + 6\boldsymbol{j} - 6\boldsymbol{k}$ acts at position $(10, 3, 4)$ relative to a coordinate system. What is the moment of the force about the origin?

13. [2.2] What is the moment of the force in Prob. 12 about the point $(6, -4, -3)$?

14. [2.2] Two forces $\boldsymbol{F}_1$ and $\boldsymbol{F}_2$ have magnitudes of 10 lb and 20 lb, respectively. $\boldsymbol{F}_1$ has a set of direction cosines $l = 0.5$, $m = 0.707$, $n = -0.5$. $\boldsymbol{F}_2$ has a set of direction cosines $l = 0$, $m = 0.6$, $n = 0.8$. If $\boldsymbol{F}_1$ acts at point $(3, 2, 2)$ and $\boldsymbol{F}_2$ acts at $(1, 0, -3)$ what is the sum of these moments about the origin?

15. [2.2] What is the moment of a 10-lb force directed along the diagonal of a cube about the corners of the cube? The side of the cube is a ft.

16. [2.2] Compute the moment of the 300-lb force shown in Fig. 2.58 about points P_1 and P_2.

17. [2.2] Find the moment of the 100-lb force shown in Fig. 2.59 about points A and B.

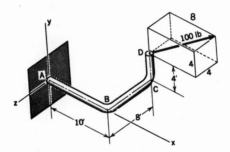

FIGURE 2.59

18. [2.2] Compute the moment of the 1000-lb force (Fig. 2.60) about supporting points A and B.

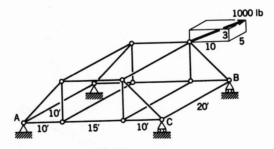

FIGURE 2.60

19. [2.2] In Fig. 2.61, cables CD and AB help support the member ED and the 1000-lb load at D. At E there is a socket joint which also sup-

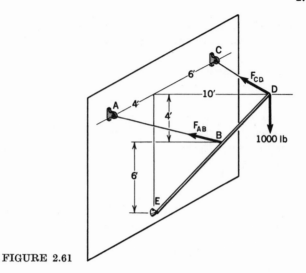

FIGURE 2.61

ports the member. Denoting the forces from the cables as F_{CD} and F_{AB} respectively, compute moments of the three forces about the point E.

20. [2.3] A force F acts at position (3, 2, 0). It is in the xy plane and is inclined at 30° from the x axis with a sense directed away from the origin. What is the moment of this force about an axis going through the points (6, 2, 5) and (0, −2, −3)?

21. [2.3] In Prob. 12 what is the moment of the force about each axis of the coordinate system? Also what is the moment of the force about the line going through the origin and having direction cosines $l = 0.5$, $m = 0.5$, $n = -0.707$? Finally, what is the moment of this force about a line parallel to the preceding one and going through the point (3, 2, 5)?

22. [2.3] Compute the thrust of the applied forces shown in Fig. 2.62 along the axis of the shaft and the torque of the forces about the axis of the shaft.

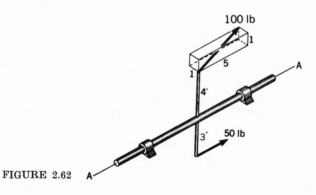

FIGURE 2.62

23. [2.3] Compute the torque from the 1000-lb force in Fig. 2.60 about the axis going through supports A and C. What is the moment about the axis between C and B?

24. [2.3] A force $F = 10i + 16j$ goes through the origin of the coordinate system. What is the moment of this force F about an axis going through points (1) and (2) with position vectors:

$$r_1 = 6i + 3k$$
$$r_2 = 16j - 4k$$

25. [2.3] A force $F = 16i + 10j - 3k$ lb goes through point (a) having a position vector $r = 16i - 3j + 12k$ ft. What is the moment about an axis going through points (1) and (2) having respective position vectors given as:

$$r_1 = 6i + 3j - 2k \text{ ft}$$
$$r_2 = 3i - 4j + 12k \text{ ft}$$

26. [2.3] A 100-lb force goes through points

$$r_1 = 10i - 3j + 12k$$
$$r_2 = 3i - 2j + 5k$$

What is the moment of such a force about a line in the xy plane going through the origin and at an angle of 30° with the x axis?

27. [2.3] In Prob. 17 find the moment of the 100-lb force about a line going from point A to point C.

28. [2.3] Find the moment of the 1000-lb force shown in Fig. 2.63 about an axis going between points D and C.

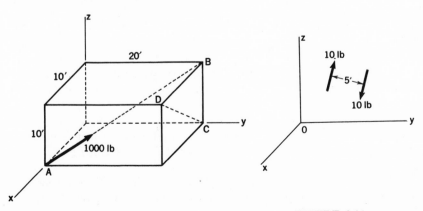

FIGURE 2.63   FIGURE 2.64

29. [2.4] A force $F_1 = 10i + 6j + 3k$ acts at position (3, 0, 2). At point (0, 2, −3) an equal but opposite force $-F_1$ acts. What is the couple-moment? What are the direction cosines of the normal to the plane of the couple?

30. [2.4] In Prob. 29, what is the perpendicular distance between the forces? If this distance is increased by 10 units, what must the magnitude of each force become to give the same couple-moment?

31. [2.5] A couple is shown in the yz plane (Fig. 2.64). What is the moment of this couple about the origin? About point (6, 3, 4)? What is the moment of the couple about a line through the origin with direction cosines $l = 0$, $m = 0.8$, $n = -0.6$? If this line is shifted to a parallel position so that it goes through point (6, 3, 4), what is then the moment of the couple about this line?

32. [2.5] Equal and opposite forces are directed along diagonals on the faces of a cube as shown in Fig. 2.65. What is the couple-moment if $a = 3$ in. and $F = 10$ lb? What is the moment of this couple about a diagonal from A to D?

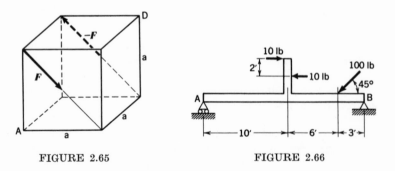

FIGURE 2.65   FIGURE 2.66

33. [2.5] Find the moment about point A of the indicated system of forces (Fig. 2.66).

34. [2.5] Add the two forces and couple shown acting on the beam in Fig. 2.67. Next, compute the moment of the forces and couple about support A.

35. [2.5] Given the two indicated force systems shown in Fig. 2.68, what is the moment of these force systems about points A and B?

36. [2.6] Shown in Fig. 2.69 are three couples. What is the total couple-moment? What is the moment of this force system about point (3, 4, 2)?

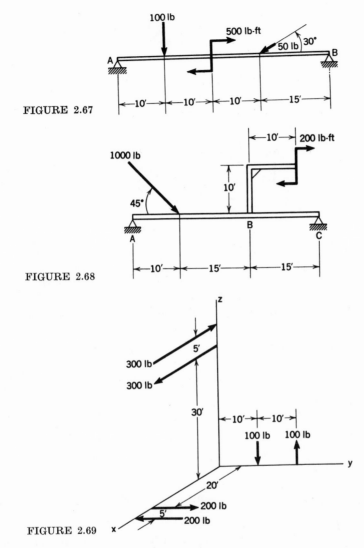

FIGURE 2.67

FIGURE 2.68

FIGURE 2.69

What is the moment of this force system about the position vector $r = 3i + 4j + 2k$? What is the total force of this system?

37. [2.6] Add a 28 lb-ft couple in the x direction and a 21 lb-ft couple in the y direction, giving the magnitude and direction of the total couple. If we wish equal and opposite forces of 5 lb, how much must they be separated to form the computed couple?

38. [2.6] Add the couples whose forces act along diagonals of the rectangular parallelepiped shown in Fig. 2.71.

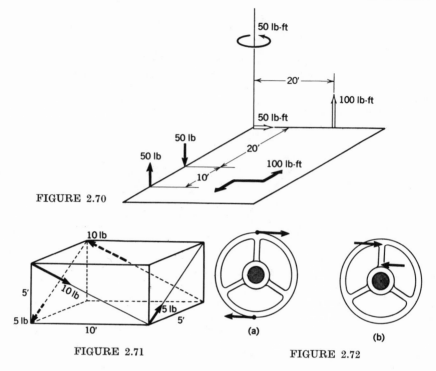

FIGURE 2.70

FIGURE 2.71   FIGURE 2.72

39. [2.6] What is the total couple-moment of the couples shown in Fig. 2.70?

40. [2.6] Equal couples in the plane of the wheel are shown in (a) and (b) of Fig. 2.72. Explain why they are equivalent for the purpose of turning the wheel. Are they equivalent from the viewpoint of the deformation of the wheel? Explain.

41. [2.7] In Fig. 2.71 find the component of the couple-moment along both diagonals.

42. [2.7] In Fig. 2.69 find the moment of the system about a line going from

$$r_1 = 6i + 10j + 3k$$

to

$$r_2 = 3i - 2j + 5k$$

43. [2.7] Given the following couple-moments:

$$C_1 = 100i + 30j + 82k \text{ lb-ft}$$
$$C_2 = -16i + 42j \text{ lb-ft}$$
$$C_3 = 15k \text{ lb-ft}$$

What couple will restrain the twisting action of this system about an axis going from

$$r_1 = 6i + 3j + 2k$$

to

$$r_2 = 10i - 2j + 3k$$

while giving a moment of 100 lb-ft about the x axis and 50 lb-ft about the y axis?

44. [2.7] Using Fig. 2.70, find the value of the maximum moment considering all possible directions where only the direction cosine $m = 0.6$ is specified. What are the other direction cosines corresponding to this maximum moment?

45. [2.9] Replace the 100-lb force in Fig. 2.73 by an equivalent system, from a rigid-body point of view, at A. Do the same for point B. Do this problem by the technique of adding equal and opposite collinear forces and also by using the cross product.

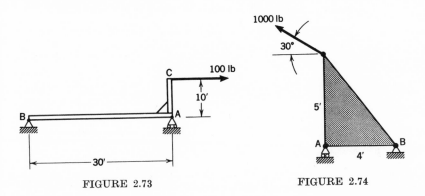

FIGURE 2.73   FIGURE 2.74

46. [2.9] In Prob. 45 explain why the new force systems would be meaningful in computing supporting forces but would be of little help in computing the deflection of point C.

47. [2.9] Replace the 1000-lb force shown in Fig. 2.74 by equivalent systems at points A and B. Do so by using the addition of equal and opposite collinear force components and by using the cross product.

48. [2.9] Replace the 500-lb force shown in Fig. 2.75 by an equivalent force system at points A and B. If a 100 lb-ft couple is added at point C as shown dashed, compute the equivalent force systems at A and B for both the 500-lb force and the couple.

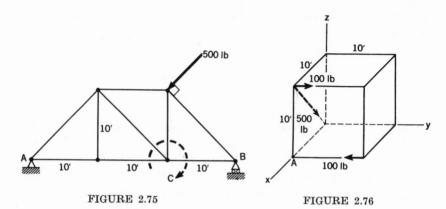

FIGURE 2.75   FIGURE 2.76

49. [2.9] A force is acting from the origin along the $+y$ axis with a magnitude of 5 lb. What is the equivalent system if we move the force to act from: (a) $(0, 0, 1)$? (b) $(0, 1, 0)$? (c) $(1, 0, 0)$? (d) $(1, 1, 1)$?

50. [2.9] A force $F = 3i - 6j + 4k$ lb goes through point $(6, 3, 2)$ ft. Replace this force by an equivalent system where the force goes through point $(2, -5, 10)$ ft.

51. [2.9] Replace the force $F = 3i + 10j + 16k$ lb, having a line of action through the origin, by an equivalent system where the F goes through the point $(2, 5, -4)$ ft.

52. [2.9] A force $F = 6i + 3j + 10k$ passes through position $(3, 4, -2)$. Replace this force by a system where the F goes through the origin.

53. [2.9] A force $F = 10i + 3j - 2k$ goes through a point whose position vector is $r = 6i - 2j$. Find an equivalent system such that the force goes through position $r = 2i + 3k$.

54. [2.9] Replace the system of forces acting on the cube in Fig. 2.76 by an equivalent system where the force goes through point A.

55. [2.9] Three forces and two couples are given as follows:

$$F_1 = 4i + 3j + 0k \text{ and acts through } (1, 0, 1)$$
$$F_2 = 5i + 5j + 5k \text{ and acts through } (1, 1, 1)$$
$$F_3 = 0i + 2j + 4k \text{ and acts through } (0, 1, 2)$$
$$C_1 = 6i + 6j + 6k$$
$$C_2 = 3i + 0j - 3k$$

Replace them with one force and couple passing through the origin.

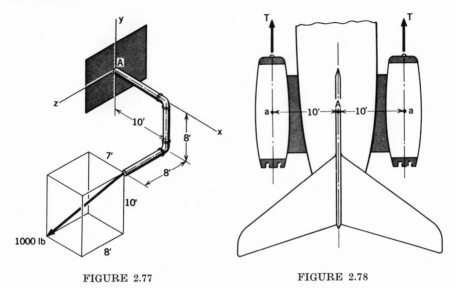

FIGURE 2.77   FIGURE 2.78

56. [2.9] What is the equivalent force system acting at point A in Fig. 2.77.

57. [2.9] Shown in Fig. 2.78 is a portion of a jet transport plane. Each jet engine is developing a thrust of 12,000 lb. If each engine weighs 2500 lb, what is the equivalent force system at position A when the plane is flying horizontally? Assume the weight of each engine acts at position a.

58. [2.9] Consider a couple C and a force F as given below.

$$C = 10k \text{ lb-ft}, \qquad F = 7i + 12j \text{ lb}$$

The force acts through the origin. To what point must we move F so that we can reduce this to an equivalent system of *one* force and no couple?

59. [2.9] Replace the forces shown in Fig. 2.79 by a single equivalent force.

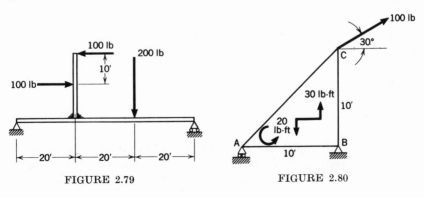

FIGURE 2.79   FIGURE 2.80

60. [2.9] Replace the force and torques shown acting on the plate in Fig. 2.80 by a single force. Give the intercept of the line of action of this force with the vertical edge of BC, the plate.

61. [2.9] Replace the forces and torques shown acting on the apparatus in Fig. 2.81 by a single force. Carefully give the line of action of this force.

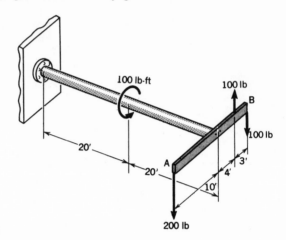

FIGURE 2.81

62. [2.10] Replace the following system of forces by a resultant system at the origin.

$$F_1 = \quad 5k \text{ and acts through } (1, 1, 0)$$
$$F_2 = -3k \text{ and acts through } (4, 2, 0)$$
$$F_3 = -2k \text{ and acts through } (-2, 3, 0)$$

63. [2.10] (a) What is the resultant of the forces in Fig. 2.82 at the origin? (b) What is the resultant of the forces at the point (0, 10, 5)?

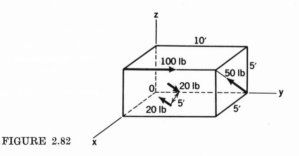

FIGURE 2.82

64. [2.10] Compute the resultant force of the applied loads shown in Fig. 2.83 at positions A and B.

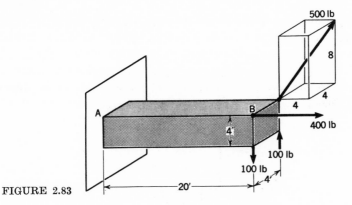

FIGURE 2.83

65. [2.10] What is the resultant of the applied loads shown in Fig. 2.84 at position *A*?

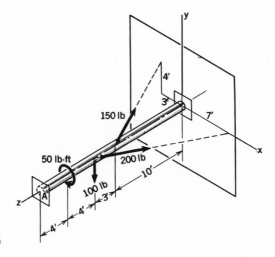

FIGURE 2.84

66. [2.10] Compute the total downward force at *A* stemming from the

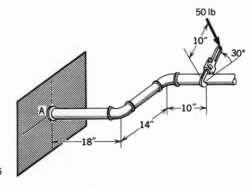

FIGURE 2.85

indicated 50-lb force in Fig. 2.85. What is the twist developed about the axis of the shaft at A?

67. [2.10] Evaluate forces F_1, F_2, and F_3 so that the resultant of the forces and torque acting on the plate in Fig. 2.86 is zero in both force and couple-moment.

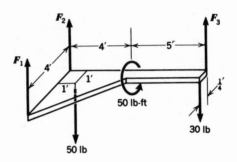

FIGURE 2.86

68. [2.10] Replace the system of forces in Fig. 2.87 by a resultant at A.

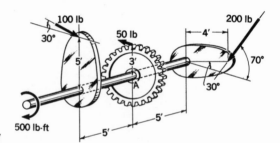

FIGURE 2.87

69. [2.10] Find the simplest resultant for the forces shown in Fig. 2.88. Give the location of this resultant clearly.

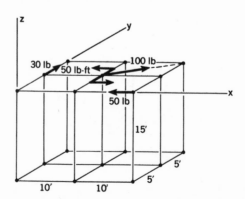

FIGURE 2.88

70. [2.10] Replace the system of forces acting on the rivets of the plate shown in Fig. 2.89 by the simplest resultant. Give the intercept of this resultant with the x axis.

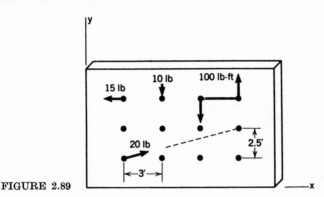

FIGURE 2.89

71. [2.10] Find the simplest resultant of the forces shown acting on the pulley in Fig. 2.90. Give the intercept with the axes.

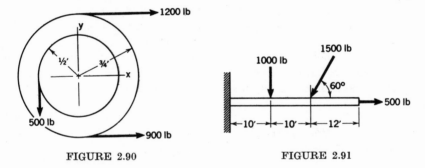

FIGURE 2.90   FIGURE 2.91

72. [2.10] Find the simplest resultant of the forces shown acting on the beam in Fig. 2.91. Give the intercept with the axis of the beam.

73. [2.10] Compute the simplest resultant for the loads shown in Fig. 2.92 acting on the beam. Give the intercept with the axis of the beam.

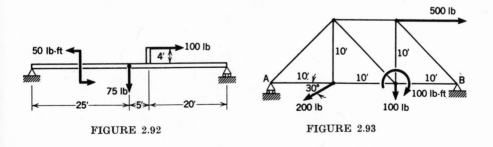

FIGURE 2.92   FIGURE 2.93

74. [2.10] What is the simplest resultant of the system of forces and torque acting on the truss shown in Fig. 2.93? Give the intercept with axis AB. What is the resultant of this system at the position of the support A?

75. [2.10] What is the simplest resultant for the force and couple acting on the beam shown in Fig. 2.94?

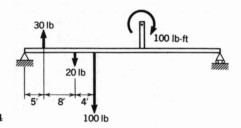

FIGURE 2.94

76. [2.10] A parallel system of forces is such that:

a 20-lb force acts at position $x = 10$, $y = 3$;

a 30-lb force acts at position $x = 5$, $y = -3$;

a 50-lb force acts at position $x = -2$, $y = 5$.

(a) If all forces point in the negative z direction, give the simplest resultant force and its line of action.
(b) If the 50-lb force points in the plus z direction, what is the simplest resultant?

77. [2.10] What is the simplest resultant force of the system shown in Fig. 2.95? The grid is composed of 1-ft squares.

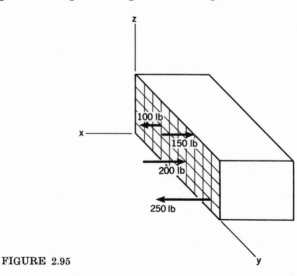

FIGURE 2.95

78. [2.10] Explain why the system shown in Fig. 2.96 can be considered a system of parallel forces. Find the simplest resultant for this system. The grid is composed of 1-ft squares.

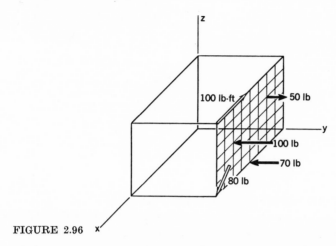

FIGURE 2.96

79. [2.10] Where should a 100-lb force in a downward direction be placed in Fig. 2.97 for the simplest resultant of all shown forces to be at position 5, 5?

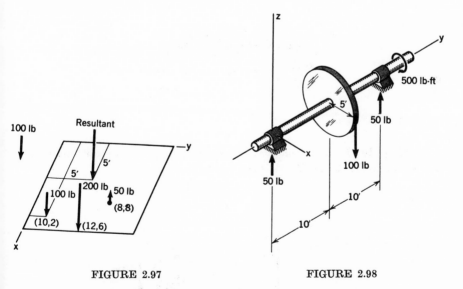

FIGURE 2.97   FIGURE 2.98

80. [2.10] What is the simplest resultant of the three forces and couple shown acting on the shaft and disc in Fig. 2.98? Disc radius is 5 ft.

81. [2.10] Compute the simplest resultant of the system of forces shown in Fig. 2.99. The grid is composed of 1-ft squares.

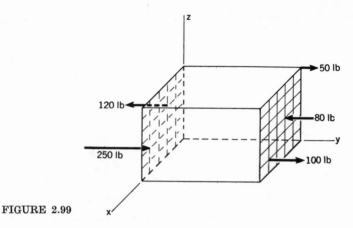

FIGURE 2.99

82. [2.10] Find the resultant at the center of gravity of a ship, at $(0, 0, 2)$ ft (Fig. 2.100), from the propeller P and rudder R. P gives a force of $1000\boldsymbol{j}$ lb which goes through $(0, -30, 0)$ ft. R gives a force of $100(\cos 30°\boldsymbol{i} + \sin 30°\boldsymbol{j})$ lb which goes through $(0, -32, 0)$ ft. What is the component of moment that helps the ship to turn?

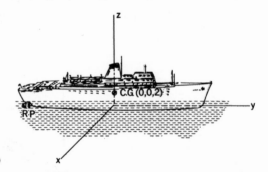

FIGURE 2.100

83. [2.10] Find the simplest resultant for the forces shown acting on the beam in Fig. 2.101. What is the resultant at position A?

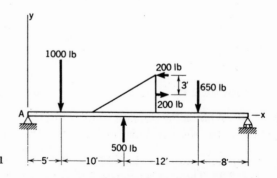

FIGURE 2.101

84. [2.11] A force field is given as:

$$F(x, y, z, t) = (10x + 5)i + (16x^2 + 2z)j + 15k$$

What is the force at position (3, 6, 7)? What is the difference between the force at this position and that at the origin?

85. [2.11] A magnetic field is developed such that the body force on the rectangular parallelepiped of metal in Fig. 2.102 is given as:

$$F = (0.01x + \tfrac{1}{8})k \text{ ounces/lbm}$$

If the specific weight of the metal is 450 lb/ft³, what is the simplest resultant body force from such a field?

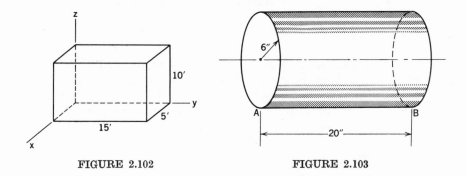

FIGURE 2.102   FIGURE 2.103

86. [2.11] In the above problem, the specific weight varies according to the relation $(450 + 2x)$ lb/ft³. Find the simplest resultant body force first from the magnetic field and then from the gravitational field.

87. [2.11] The specific weight γ of the material in the solid cylinder shown in Fig. 2.103 varies linearly as one goes from face A to face B. If

$$\gamma_A = 400 \text{ lbf/ft}^3, \qquad \gamma_B = 500 \text{ lbf/ft}^3$$

what is the position of the center of gravity of the cylinder?

88. [2.11] Shown in Fig. 2.104 is a hollow cylinder made up of three portions, each portion having a constant specific weight. What is the center of gravity for the cylinder?

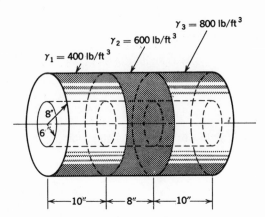

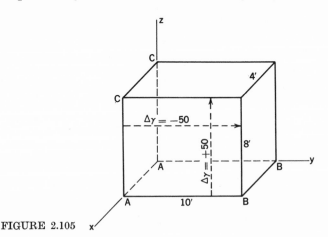

FIGURE 2.104

89. [2.11] Shown in Fig. 2.105 is a block of material whose specific weight varies in both the y and z directions. Along the x axis (at AA) the specific weight is 200 lb/ft^3, decreasing in value linearly by 50 lb/ft^3 as one goes to the other end at BB. The same decrease takes place at all constant elevations z. However, as elevation z is changed for any value y the specific weight increases linearly by 50 lb/ft^3 as one goes from the bottom face to the upper face. The dashed arrows indicate these changes.

(a) Express γ as a function of x, y, z.

(b) Compute the xy coordinates of the center of gravity.

FIGURE 2.105

90 [2.11] Part of the block in Prob. 89 has been cut away as indicated in Fig. 2.106. Using the data of Prob. 89, compute coordinates xy of the center of gravity of the material left.

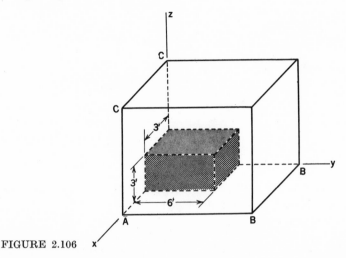

FIGURE 2.106

91. [2.11] The specific weight of the material in a right circular cone, Fig. 2.107, varies directly as the square of the distance y from the base. If γ_0 is the specific weight at the base and γ' is the specific weight at the tip, what is the center of gravity of the cone?

92. [2.11] In Prob. 91, suppose a cylindrical element is cut from the cone as shown in Fig. 2.108. What is the new position of the center of gravity? This cylinder has a radius a and a height b.

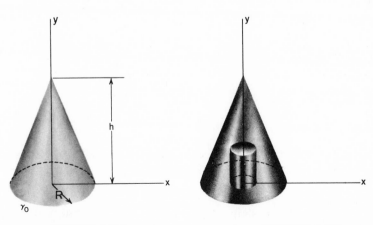

FIGURE 2.107   FIGURE 2.108

93. [2.11] A triangular plate of thickness t, height a, and base b is shown in Fig. 2.109. What is the center of gravity of this plate?

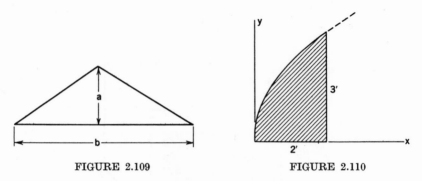

FIGURE 2.109   FIGURE 2.110

94. [2.11] Shown in Fig. 2.110 is a plate of thickness t. The upper edge has a parabolic shape with infinite slope at the origin. Find the coordinates of the center of gravity for this plate.

95 [2.11] Shown in Fig. 2.111 is a plate 6 inches thick. The curved edge is that of a parabola with zero slope at the origin. Find the coordinates of the center of gravity.

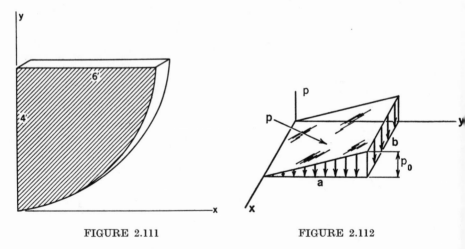

FIGURE 2.111   FIGURE 2.112

96. [2.11] Find the resultant of a normal pressure distribution over the rectangular area with sides a and b in Fig. 2.112. Give the center of pressure.

97. [2.11] Find the simplest resultant acting on wall $ABCD$, Fig. 2.113. Give the coordinates of the center of pressure. The pressure varies such that $p = A/(y + 1) + B$ from 10 psi to 50 psi as indicated in the diagram.

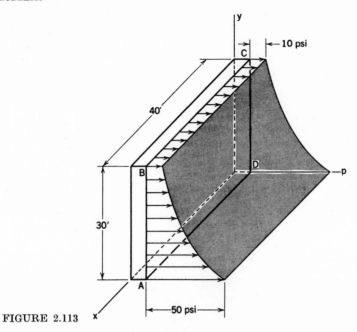

FIGURE 2.113

98. [2.11] Find the simplest resultant force for the pressure distribution shown acting on the block in Fig. 2.114. Give the coordinates of the centers of pressure. The pressure on the upper surface varies parabolically. The pressure on the lower surface varies linearly.

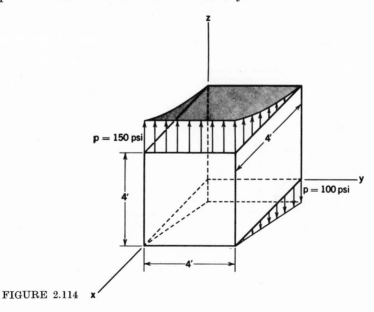

FIGURE 2.114

99. [2.11] Shown in Fig. 2.115 is a block 1 ft thick submerged in water. You will learn that pressure increases linearly with depth according to the formula $p = \gamma d$ where γ is the specific weight of water and d is the depth below the surface of the water. Compute the simplest resultant force and the center of pressure on the surface shown on edge as AB. Take $\gamma = 62.4$ lb/ft^3.

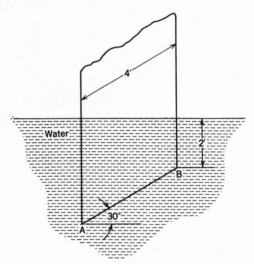

FIGURE 2.115

100. [2.11] Water is contained in the glass rectangular tank shown in Fig. 2.116. A triangular block of wood is shown submerged in the water. The edge AB is parallel to the surface of the water at a distance 6 ft below. What is the simplest resultant from the water pressure on the total closed surface of the triangular block? (See Prob. 99 before doing.)

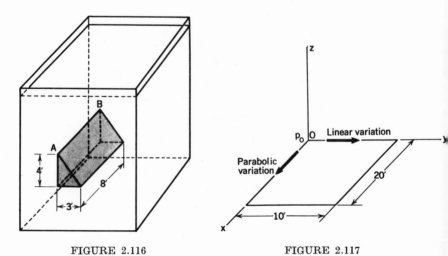

FIGURE 2.116   FIGURE 2.117

101. [2.11] The pressure p_0 at the corner O of the plate shown in Fig. 2.117 is 50 psi and increases linearly in the y direction by 5 lb/ft²/ft. In the x direction it increases parabolically starting with zero slope so that in 10 ft the pressure has gone from 50 psi to 500 psi. What is the simplest resultant for this distribution? Give the coordinates of the center of pressure.

102. [2.11] In Fig. 2.118 is shown a pressure distribution p forming a hemispherical surface over a domain of radius 5 ft. If the maximum pressure is 5 lb/ft², what is the resultant from this pressure distribution?

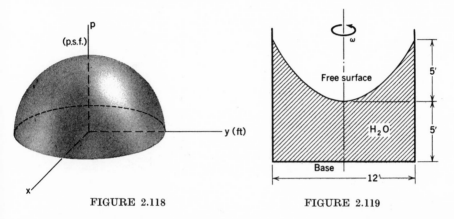

FIGURE 2.118   FIGURE 2.119

103. [2.11] Suppose in Prob. 102 that the domain had a radius of 15 ft and that the scale of measure of pressure were such that the pressure surface was still a hemisphere with a maximum value of 5 lb/ft². What would be the resultant force?

104. [2.11] A cylindrical tank of water is rotated at constant angular speed ω until the water ceases to change shape. The result (Fig. 2.119) is a free surface which, from fluid mechanics considerations, is that of a paraboloid. If the pressure varies directly as the depth below the free surface, what is the resultant force on a quadrant of the base of the cylinder?

105. [2.11] The sail ABC of a sailboat has a shape as shown in Fig. 2.120 (side view), where AB is a portion of a circle and BC is a straight line. If the wind has an effect equivalent to a pressure of $p = 10$ lb/ft² acting on the rear of the sail, what is the component of force that pushes the sailboat ahead? What is the resultant at the center of gravity D? Assume the sail has a width of 5 ft.

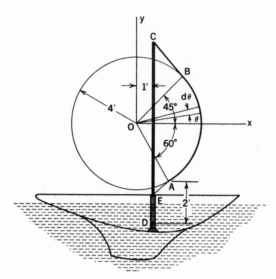

FIGURE 2.120

106. [2.11] Find the simplest resultant of the forces shown acting on the beam in Fig. 2.121.

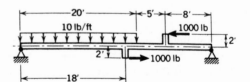

FIGURE 2.121

107. [2.11] We have a parabolic load and a point load acting on the beam as shown in Fig. 2.122. Find the simplest resultant for this system of loads. Give the line of action.

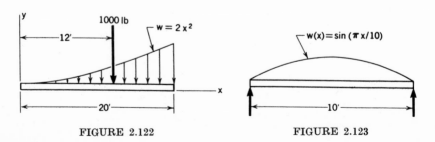

FIGURE 2.122   FIGURE 2.123

108. [2.11] Replace the continuous load in Fig. 2.123 by the simplest equivalent system. If we require the resultant force to go through the right end of the beam, what is the equivalent system?

109. [2.11] Compute the simplest resultant for the loads shown acting on the simply supported beam in Fig. 2.124. Give the line of action.

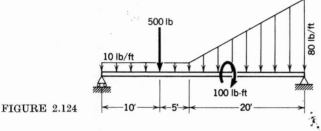

FIGURE 2.124

110. [2.11] Compute the simplest resultant force for the loads shown acting on the cantilever beam in Fig. 2.125. What force and moment are transmitted by this force to the supporting wall at A?

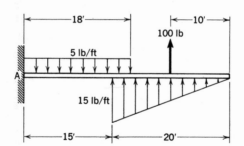

FIGURE 2.125

111. [2.11] Shown in Fig. 2.126 is wire $ABCD$. The weight of the wire per unit length w increases linearly from 4 ounces per foot at A to 20 ounces per foot at D. What is the center of gravity of the wire?

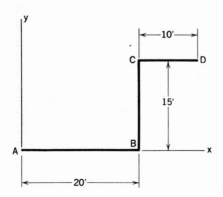

FIGURE 2.126

112. [2.11] Find the center of gravity of the wire shown in Fig. 2.127. The weight per unit length increases as the square of the length of wire from a value of 3 ounces per foot at A until it reaches the value of 8 ounces per foot at C. It then decreases 1 ounce per foot for every ten feet of length.

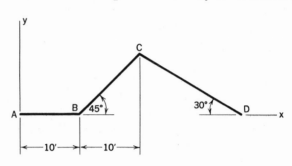

FIGURE 2.127

Chapter 3

EQUILIBRIUM EQUATIONS

3.1. Free-Body Diagram

You will recall from Section 1.5 that a body in equilibrium is one that is stationary or one wherein each point moves uniformly relative to an inertial reference. When taken as a rigid body, there are then certain simple equations that relate all the surface and body forces, or their equivalents, acting on such a body. With these equations we can sometimes ascertain the value of a certain number of unknown forces. For instance, in the beam shown in Fig. 3.1 we know the loads F_1 and F_2 and their positions in the undeformed geometry as well as the weight of the beam. We want to determine the forces transmitted to the earth so we can design a foundation to support the structure properly. Knowing that the beam is in equilibrium and that the deformation of the beam is small enough so that we can use the original undeformed geometry, we may relate known and unknown forces via the rigid-body equations of equilibrium and thereby arrive at the desired information.

Since these equations actually stem from the dynamical considerations of a body, we must be sure to include *all* the forces (or their equivalents) acting *on* this body, because they all affect the motion of

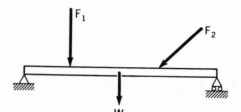

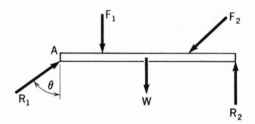

FIGURE 3.1

FIGURE 3.2

a body and must be accounted for. To help identify all the forces and so insure the correct use of the equations of statics, we isolate the body in a simple diagram and show *all* the forces from the *surroundings* that act *on* the body. Such a diagram is called a *free-body diagram*. When we isolate the beam in our problem from its surroundings, we get Fig. 3.2. On the left end, there is an unknown force from the ground that has a magnitude denoted as R_1 and a direction denoted as θ with a line of action going through a known point A. [We may also use components $(R_1)_x$ and $(R_1)_y$ as unknowns rather than R_1 and θ.] The right side involves a force in the vertical direction with an unknown magnitude denoted as R_2. The direction is vertical because the beam is on rollers to allow for thermal expansion, and the ground, therefore, exerts a negligibly small horizontal force on the beam. Once all the forces acting on the beam have been identified, including the three unknown quantities R_1, R_2, and θ, we can by using three scalar equations of equilibrium, solve for these unknowns.

Consider now the two spheres shown in Fig. 3.3 in a condition of equilibrium, with surfaces

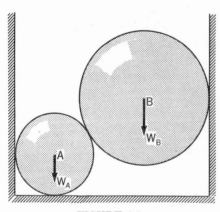

FIGURE 3.3

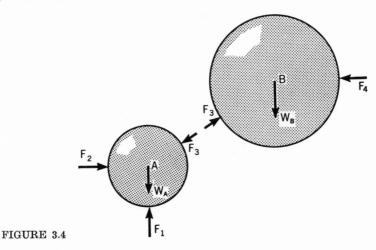

FIGURE 3.4

smooth and hard enough to permit us to neglect friction completely. The contact forces thus must be in a direction normal to the surface of contact. The free bodies of the spheres are shown in Fig. 3.4. Notice that F_3 is the magnitude of the force from sphere B on sphere A, while the reaction, alos shown as F_3, is the magnitude of the force from sphere A on sphere B.

You might be tempted to consider a portion of the container as a free body in the manner shown in Fig. 3.5, but even if this diagram did clearly depict a body (which it does not) it would not qualify as a free body, since all the forces acting on the body have not been shown.

In engineering problems, bodies are often in contact in a number of standard ways. In Fig. 3.6 you will find the types of forces transmitted from body M to body N for body connections that are often

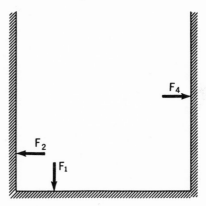

FIGURE 3.5

found in practice. These are not free-body diagrams, since all the forces on any body have not been shown.

In general, to ascertain the nature of the force system that a body M is capable of transmitting to a second body N through some connector or support we may proceed in the following manner. Mentally move the bodies relative to each other in each of three orthogonal directions. In those directions where relative motion is impeded or prevented

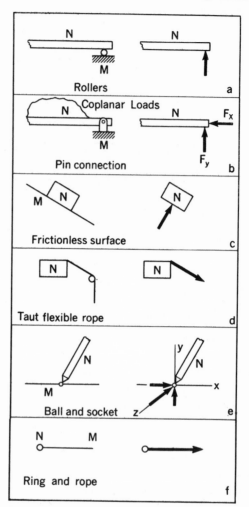

FIGURE 3.6

by the connector or support there will be a force component at this
connector or support in a free-body diagram of either body M or N.
Next mentally rotate bodies M and N relative to each other about the
orthogonal axes. In each direction about which relative rotation is
impeded or prevented by the connector or support there will be a
couple-moment component at this connector or support in a free-body
diagram of body M or N. Now it may turn out as a result of equilibrium
considerations of body M or N that certain of these force and moment
components that are capable of being generated at a support or connector are zero for the particular loadings at hand. Indeed, one can
often recognize this by inspection. In Fig. 3.6(b) we have done this

very thing. We have tacitly assumed that the loadings on N were coplanar in the plane of the page. Consequently, while the pin connection is capable of transmitting components F_z, M_y, and M_x, these have been taken as zero and not shown because we know that they must be zero for such loadings. You are urged to examine the other cases using the suggested mental exercises.

3.2. Free Bodies Involving Interior Sections

Let us consider a rigid body in equilibrium as shown in Fig. 3.7. Clearly, every portion of this body must also be in equilibrium. If we consider the body as two parts A and B we can present either part in a free-body diagram. To do this, we must include on the part chosen to be the free body the forces from the other part that arise at the common section (Fig. 3.8). The surface between both parts may be any curved surface, and over it there will be a continuous force distribution. In the general case, we know that such a distribution can be replaced by a single force and a couple, and this has been done in the free-body diagram of parts A and B in Fig. 3.8.

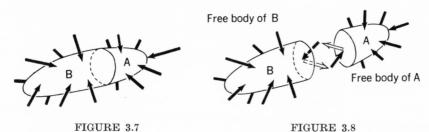

FIGURE 3.7   FIGURE 3.8

This technique of "exposing" the interior of a body for analysis is most useful. We can often solve the equivalent system for the section, given by the force and couple, by using equations of equilibrium. By employing other laws—Hooke's law, for example—we can, as in the case of strength of materials, often determine the distribution of force.

As an example, consider a beam with one end embedded in a massive wall and loaded along the center plane (Fig. 3.9). A free body of the portion of the beam extending from the wall is shown in Fig. 3.10. Because of the geometric symmetry about the center plane, and the fact that the loads are in this plane, the exposed forces in the cut section can be considered coplanar. Hence this distribution can be replaced by a force and a couple in the center plane. Although a line of

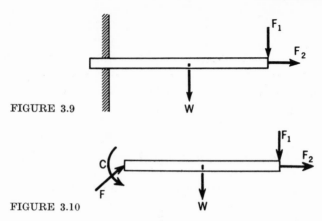

FIGURE 3.9

FIGURE 3.10

action for the force can be found that would enable us to eliminate the couple, it is desirable in structural problems to work with an equivalent system that has the force passing through the center of the beam cross section, and thus to have a couple. In the next section we will see how F and C may be ascertained.

EXAMPLE 3.1

As a further illustration of a free-body diagram, we shall now consider the assembly shown in Fig. 3.11, which consists of members connected by frictionless pins. The force systems acting on the assembly and its parts will be taken as coplanar. We shall now set forth several free-body diagrams of the assembly and its parts.

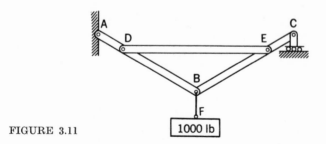

FIGURE 3.11

a. Free-body diagram of the entire assembly. The magnitude and direction of the force at A from the wall onto the assembly are not known. However, we know that this force is in the plane of the system. Therefore, two components are shown at this point (Fig. 3.12). Since the direction of the force C is known, there are then three unknown scalar quantities, A_y, A_x, and C for the free body.

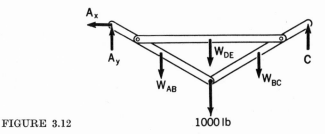

FIGURE 3.12

b. Free-body diagram of the component parts. When two members are pinned together, such as members *DE* and *AB* or *DE* and *BC*, we usually consider the pin to be part of one of the bodies, but when more than two members are connected at a pin, such as members *AB*, *BC*, and cable *BF* at *B*, we often isolate the pin and consider that all members act on the pin rather than directly on each other, as is illustrated in Fig. 3.13. Notice the forces that form pairs of reactions have been associated with dotted enclosures.

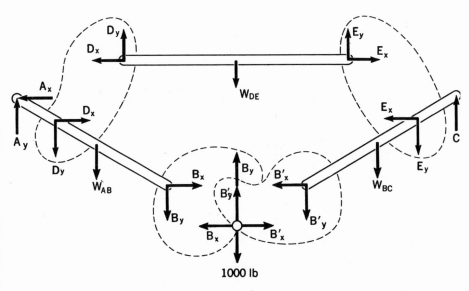

FIGURE 3.13

Do not be concerned about the proper sense of an unknown force component that you enter on the free-body diagram, for you may choose either a positive or negative sense for these components. When the values of these quantities are ascertained by methods of statics, the proper sense for each component can then be established. Having chosen a sense for a component, you must, however, be sure that the reaction to this component has the opposite sense.

c. Free-body diagram of portion of the assembly to the right of MM.
In making a free body of the portion to the right of section MM (see Fig. 3.14), we must remember to put in the weight of the portions of the member remaining after the cut has been made. Notice in Fig. 3.15 that there are seven unknown scalar quantities for this free-body diagram. They are C, C_1, C_2, F_{1x}, F_{1y}, F_{2x}, and F_{2y}. It is apparent from this problem that the number of unknowns varies widely for the various free bodies that may be drawn in the system. For this reason you must choose the free-body diagram that is suitable for your needs with some discretion in order to solve effectively the desired unknowns.

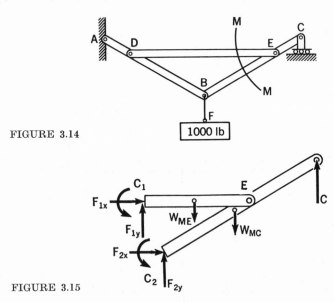

FIGURE 3.14

FIGURE 3.15

3.3. Equations of Equilibrium

In every free-body diagram it is possible to replace the system of forces and couples acting on the body by a single force and a single couple acting on the body at some point a. The force will have the same magnitude and direction, no matter what point a is chosen to move the entire system, by methods discussed earlier. However, the couple will depend on the chosen point. It can be proved in dynamics that *the necessary and suffcient conditions for a rigid body to be in equilibrium are that the resultant force F_R and the resultant couple C_R for any point a be zero vectors.* That is:

$$F_R = 0 \qquad \text{(a)}$$
$$C_R = 0 \qquad \text{(b)}$$

3.1

These may be considered the fundamental equations of statics. You will remember from Section 2.10 that the resultant F_R is the sum of the forces moved to the common point and the couple C_R is equal to the sum of the moments of all the original forces taken about this point plus all the couple moments. Hence the above equation can be written as:

$$\sum_{i=1}^{n} F_i = 0 \qquad \text{(a)}$$
$$\sum_{i=1}^{n} \rho_i \times F_i + \sum_{i=1}^{m} C_i = 0 \qquad \text{(b)}$$

3.2

where the ρ_i's are displacement vectors from the common point to any point on the lines of action of the respective forces. From this form of the equations of statics, we can conclude that for equilibrium to exist *the vector sum of the forces must be a zero vector and the moment of the system of forces and couples about any point in space must be a zero vector.*

Now that we have summed forces and have taken moments about a point a, we will demonstrate that we cannot find another independent equation by taking moments about a different point b. For the body in Fig. 3.16 shown in equilibrium, we have initially the following equations of equilibrium:

$$F_1 + F_2 + F_3 + F_4 = 0 \qquad \qquad \textbf{3.3}$$
$$\rho_1 \times F_1 + \rho_2 \times F_2 + \rho_3 \times F_3 + \rho_4 \times F_4 = 0 \qquad \textbf{3.4}$$

The new point b is separated from a by the displacement vector d.

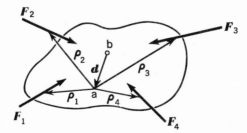

FIGURE 3.16

The displacement vectors from b to the lines of action of the forces can be given in terms of d and the displacement vectors employed in Eq. 3.4, and can be expressed in the following manner:

$$(\boldsymbol{\rho}_1)_b = (\boldsymbol{d} + \boldsymbol{\rho}_1)$$
$$(\boldsymbol{\rho}_2)_b = (\boldsymbol{d} + \boldsymbol{\rho}_2), \text{ etc.}$$

The moment equation for point b then may be given as:

$$(\boldsymbol{\rho}_1 + \boldsymbol{d}) \times \boldsymbol{F}_1 + (\boldsymbol{\rho}_2 + \boldsymbol{d}) \times \boldsymbol{F}_2 + (\boldsymbol{\rho}_3 + \boldsymbol{d}) \times \boldsymbol{F}_3 + (\boldsymbol{\rho}_4 + \boldsymbol{d}) \times \boldsymbol{F}_4 = 0 \quad \textbf{3.5}$$

Using the distributive rule for cross products, we can restate this equation as:

$$(\boldsymbol{\rho}_1 \times \boldsymbol{F}_1 + \boldsymbol{\rho}_2 \times \boldsymbol{F}_2 + \boldsymbol{\rho}_3 \times \boldsymbol{F}_3 + \boldsymbol{\rho}_4 \times \boldsymbol{F}_4) + \boldsymbol{d} \times (\boldsymbol{F}_1 + \boldsymbol{F}_2 + \boldsymbol{F}_3 + \boldsymbol{F}_4) = 0$$
$$\textbf{3.6}$$

Since the second parenthesis is zero in accordance with Eq. 3.3, the remaining portion degenerates to Eq. 3.4, and thus we have not introduced a new equation. Therefore, there are only two independent vector equations of equilibrium for any single free body.

We will leave it to you to show, in considering equilibrium for the body examined in the preceding paragraph, that the pair of Eqs. 3.4 and 3.5 are generally equivalent to the pair of Eqs. 3.3 and 3.4. This means that we can usually use moments about two points in forming the basic equations of equilibrium and thus do not have to sum the forces and then take moments.

Using the vector Eqs. 3.2, we can now express the scalar equations of equilibrium. Since, as you will recall, the scalar component of the moment of a force about a point is the moment of the force about an axis through the point, we may state these equations in the following manner:

$\sum_i (F_x)_i = 0$ (a)	$\sum_i (M_x)_i = 0$ (d)	
$\sum_i (F_y)_i = 0$ (b)	$\sum_i (M_y)_i = 0$ (e)	**3.7**
$\sum_i (F_z)_i = 0$ (c)	$\sum_i (M_z)_i = 0$ (f)	

where $(M_x)_i$, $(M_y)_i$, and $(M_z)_i$ are moments of a force or a couple about the respective x, y, and z axes through the point. From this set of equations, it is clear that no more than *six unknown scalar quantities in the general case can be solved by methods of statics for a single free body.*

It is quite a simple matter for a given free body to express many scalar equations of equilibrium by selecting references that have different directions, in which we can sum forces, and different axes, about which we can take moments. However, in choosing six independent equations, we will find that the remaining equations will be dependent on these six, i.e., the remaining equations will be sums, differences, etc. of the independent set and so will be of no use in solving for desired unknowns.

3.4. Special Cases of Equilibrium

We shall now consider a number of important special cases of equilibrium primarily to ascertain the number of scalar equations that are necessary and sufficient for equilibrium. With this information, we will then know the number of unknown scalar quantities for any free body that can be solved by methods of statics. If there are more unknowns than there are available equations, no amount of algebraic perseverance will lead to the solution of the unknowns for the chosen free body.

The simplest type of resultant for each special system of forces is most useful in determining the number of scalar equations available in a given problem. The procedure is to classify the force system, establish what simplest resultant force system is associated with the classification, and then consider the number of scalar equations necessary to guarantee this resultant to be zero. The following cases exemplify this procedure.

CASE A. CONCURRENT SYSTEM OF FORCES. In this case, since the resultant is a single force at the point of concurrency, it is only necessary for equilibrium that this force be zero. We can insure this condition if the orthogonal components of this force are separately equal to zero. Thus we have three equations of equilibrium of the form:

$$\sum_i (F_x)_i = 0, \qquad \sum_i (F_y)_i = 0, \qquad \sum_i (F_z)_i = 0 \qquad \textbf{3.8}$$

As was pointed out in the general vector discussion, there are other ways of insuring a zero resultant here. For instance, the fact that the moment of this force system about some axis α (see Fig. 3.17) is zero may mean one of the following things:

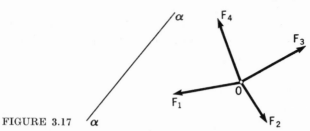

FIGURE 3.17

a. The resultant force is zero.
b. The resultant force has a line of action through the axis.
c. The resultant force has a line of action parallel to the axis.

If another nonparallel axis β is chosen, and the moment about this axis is also zero, it means that either:

a. The resultant force is zero, or
b. The resultant force has a line of action through both axes.

The previous condition c cannot be extended to the second axis, since a force cannot be parallel to two nonparallel axes simultaneously.

A third axis γ is chosen so that no line can intersect all three axes or intersect any two axes while being parallel to the third. If in addition to being zero for axes α and β the moment of the force system is zero about the axis γ, then:

The resultant force is zero.

Thus we have another way to guarantee equilibrium for a concurrent force system:

$$\sum_i (M_\alpha)_i = 0, \qquad \sum_i (M_\beta)_i = 0, \qquad \sum_i (M_\gamma)_i = 0 \qquad \textbf{3.9}$$

where axes α, β, and γ have the restrictions we have already mentioned.

We can show other sets of equations for expressing the condition of equilibrium.*

---

*For instance:

1. If the summation of forces in the y direction is zero, i.e., $\sum (F_y)_i = 0$, and
2. If the moments are zero about two axes d and e which are not parallel to the xz plane and which are so oriented that the line of action of the resultant force does not intersect both of them,

then the body on which the system of forces acts is in equilibrium. (We urge you to reason this out yourself.)

The essential conclusion to be drawn is that there are three indepen- dent equations of equilibrium for a concurrent force system. In employing scalar equations of equilibrium directly, we need not be too concerned about the choice of direction for summation of forces or the choice of axis about which to take moments. If we employ axes that violate restrictions such as those given above, we will get equations that are not independent, and we must then continue to choose other axes until three independent equations are finally secured.

CASE B. COPLANAR FORCE SYSTEM. In this case we have shown that the system may be simplified into a resultant of a single force or a single couple. Thus, to insure that there is a zero resultant force we require for a coplanar system in the xy plane:

$$\sum_i (F_x)_i = 0, \qquad \sum_i (F_y)_i = 0 \qquad \textbf{3.10}$$

And to insure that there is a zero resultant couple, we require for mo- ments about an axis α inclined to the plane:

$$\sum_i (M_\alpha)_i = 0 \qquad \textbf{3.11}$$

We conclude that *there are three scalar equations of equilibrium for a coplanar force system.* Other combinations, such as two moments and a single summation, may be employed to give the three necessary, in- dependent scalar equations of equilibrium, as was discussed in the previous case.

CASE C. PARALLEL FORCES IN SPACE. In the case of parallel forces in space, we already know that the resultant could be either a single force or a couple. If the z direction corresponds to the direction of the forces, to insure that there be a zero resultant force it is necessary that:

$$\sum_i (F_z)_i = 0 \qquad \textbf{3.12}$$

And to guarantee that there be a zero resultant couple, we require that:

$$\sum_i (M_x)_i = 0, \qquad \sum_i (M_y)_i = 0 \qquad \textbf{3.13}$$

where the x and y axes are chosen in any plane perpendicular to the direction of the forces. *Thus three independent scalar equations are available for equilibrium of parallel forces in space.*

A summary of the special cases discussed in this section is given below. It will be left for you to make similar conclusions for even simpler systems such as the concurrent-coplanar and the parallel-coplanar systems.

SUMMARY FOR SPECIAL CASES

System	Simplest resultant	Number of equations for equilibrium
Concurrent (3D)	Single force	3
Coplanar	Single force or single couple	3
Parallel (3D)	Single force or single couple	3

3.5. Problems of Equilibrium

We shall now examine problems of equilibrium in which the rigid-body assumption is valid. To solve such problems, we usually must ascertain the value of certain unknown forces. To involve these unknown forces in equations, we must draw a free-body diagram of the entire system or a portion of it to expose them for analysis. As we have seen, for any free body there are a limited number of independent scalar equations, so that at times we must employ several free-body diagrams for portions of the system to produce enough equations to solve all the unknowns.

For any free body we may proceed by expressing two basic vector equations of statics. After carrying out vector operations such as cross products, additions, etc. in the equations, we form the scalar equations which are then solved simultaneously to give scalar quantities associated with the unknown forces. We can also express the scalar equations immediately by using the alternate scalar equilibrium relations we formulated above. In the first case, we start with more compact vector equations and arrive at the expanded scalar equations by the formal procedures of vector algebra, whereas in the latter case we evaluate the expanded scalar equations by carrying out arithmetic operations on the free-body diagram as we write the equations. Which procedure is more desirable? It all depends on the problem and the investigator's skill in vector manipulation. It is true that most statics problems submit easily to a direct scalar approach, but the more challenging problems of dynamics definitely favor an initial vector approach. In this text we shall employ both procedures as the occasion warrants. Admittedly we shall at times use the vector approach where it may be somewhat clumsy (because it is too "powerful" for the job) so that we can develop a working familiarity with the vector computations necessary for dynamics.

In statics problems, we must occasionally assign a sense to a component of an unknown vector or couple in order to write the equations. If on solving the equations we obtain a negative sign for this component, it means that we have guessed the wrong sense and must reverse it in the statements of the solution. We shall now solve and discuss a number of problems of equilibrium.

EXAMPLE 3.2

A 500-lb weight is suspended by flexible cables as is shown in Fig. 3.18. Determine the tension in the cables.

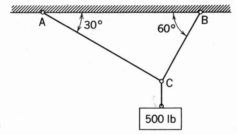

FIGURE 3.18

A suitable free body that exposes the desired unknown quantities is the ring C, which may be considered as a particle for this computation because of its comparatively small size (Fig. 3.19). The force system acting on a particle must always be a concurrent system. Here we have the additional fact that it is coplanar as well, and therefore we may solve for the two unknowns. We shall proceed directly to the scalar equations of equilibrium. Thus:

$$\underline{\Sigma\, F_y = 0}$$

$$-500 + T_{CB} \sin 60° + T_{AC} \sin 30° = 0 \qquad \text{(a)}$$

$$\underline{\Sigma\, F_x = 0}$$

$$T_{AC} \cos 30° = T_{CB} \cos 60° \qquad \text{(b)}$$

By solving these equations simultaneously, we get the desired results:

$$T_{CB} = 433 \text{ lb}, \qquad T_{AC} = 250 \text{ lb}$$

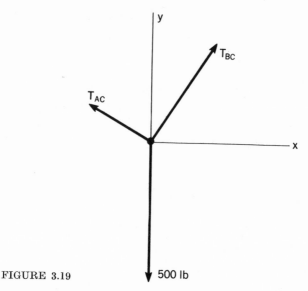

FIGURE 3.19   ↓ 500 lb

Another way of arriving at this answer is to consider the force polygon (see appendix) for the system of forces. Because the forces are in equilibrium, the polygon must close, i.e., the head of the final force must coincide with the tail of the initial force. In this case we have a right triangle, as is shown in Fig. 3.20, drawn approximately to scale.

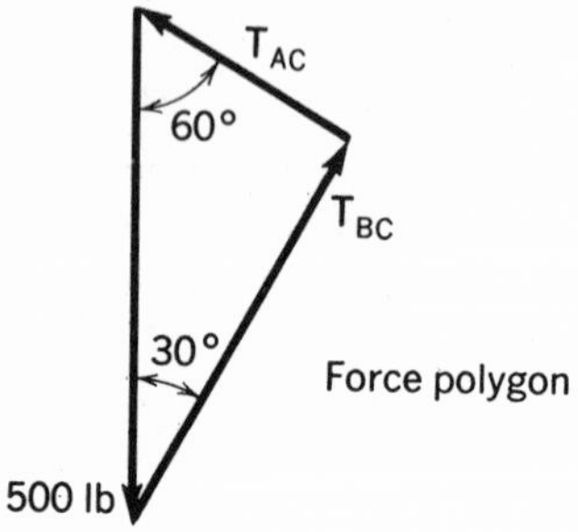

From trigonometric considerations of this right triangle, we can state:

$$T_{BC} = 500 \cos 30° = 433 \text{ lb}$$

$$T_{AC} = 500 \sin 30° = 250 \text{ lb}$$

The force polygon may be used to good advantage when three concurrent forces are in equilibrium.

Let us now initiate the computations for the unknown tensions directly from the basic equations of statics. First we must express all forces in vector notation:

FIGURE 3.20

$$\boldsymbol{T}_{CB} = T_{CB}(0.500\boldsymbol{i} + 0.866\boldsymbol{j})$$

$$\boldsymbol{T}_{AC} = T_{AC}(-0.866\boldsymbol{i} + 0.500\boldsymbol{j})$$

We get the following equation when the vector sum of the forces is set equal to zero:

$$T_{CB}(0.500\boldsymbol{i} + 0.866\boldsymbol{j}) + T_{AC}(-0.866\boldsymbol{i} + 0.500\boldsymbol{j}) - 500\boldsymbol{j} = \boldsymbol{0}$$

Choosing point C, the point of concurrency, we see clearly that the sum moments of the forces about this point are zero, so the second basic equation of equilibrium is intrinsically satisfied. We now regroup the above equation in the following manner:

$$(0.500T_{CB} - 0.866T_{AC})\boldsymbol{i} + (0.866T_{CB} + 0.500T_{AC} - 500)\boldsymbol{j} = \boldsymbol{0}$$

To satisfy this equation, each of the quantities in parentheses must be zero. This gives the scalar equations (a) and (b) stated earlier from which the scalar quantities T_{CB} and T_{AC} can be solved.

EXAMPLE 3.3

A crane weighing 300 lb supports a 10,000-lb load as is shown in Fig. 3.21. Determine the supporting forces at A, which is a pinned connection, and at B, which is a roller.

A free-body diagram of the main structure exposes the desired unknowns (Fig. 3.22). Note that since the system of forces may be taken as coplanar, we may solve for the three unknown quantities from this single free-body diagram. Equating the sum of the force vectors to zero, we get:

$$A_x\boldsymbol{i} + A_y\boldsymbol{j} + B\boldsymbol{i} - 3000\boldsymbol{j} - 10,000\boldsymbol{j} = \boldsymbol{0} \qquad \textbf{(a)}$$

Taking moments about point A, we have:

$$[-(5)(3000) - (15)(10,000) + (5)(B)]\boldsymbol{k} = \boldsymbol{0} \qquad \textbf{(b)}$$

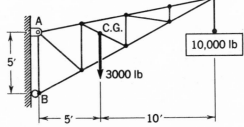

FIGURE 3.21

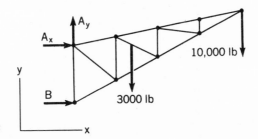

FIGURE 3.22

These equations yield the following scalar equations:

$$A_x + B = 0, \quad A_y - 13{,}000 = 0, \quad 5B - 165{,}000 = 0$$

These are easily solved to give:

$$B = 33{,}000 \text{ lb}, \quad A_x = -33{,}000 \text{ lb}, \quad A_y = 13{,}000 \text{ lb}$$

Note that a negative sign is present for A_x. This indicates that the wrong sense was chosen for this component at the outset of the computations, so we must reverse it.

We have now solved the forces from the wall onto the structure. The forces from the structure onto the wall are the reactions to these forces.

EXAMPLE 3.4

Shown in Fig. 3.23 are members AB and BC of lengths 10 ft and 7 ft respectively supporting identical cylinders each weighing 1000 lb. If the surfaces of contact of the cylinders are frictionless, compute the supporting forces at pins A and C. Assume furthermore that pins A, B, and C are frictionless and that the weights of the members AB and BC are negligible.

A free-body diagram exposing the desired unknown forces is shown in Fig. 3.24. We have here a coplanar system of forces for which three equations are available. However, we have six unknown forces, so we must use other free bodies. Accordingly, we consider in Fig. 3.25 free-body diagrams of the cylinders. These are coplanar concurrent forces systems for which we can determine forces D, G, E, and H as follows:

Cylinder J	*Cylinder K*
$\sum F_y = 0$	$\sum F_y = 0$
$G \cos 45° = 1000$	$H \sin 60° = 1000$
$G = 1414 \text{ lb}$	$H = 1155 \text{ lb}$

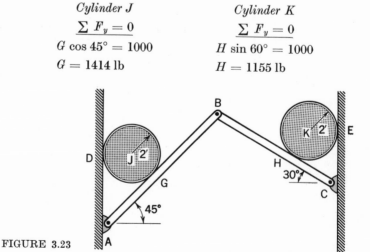

FIGURE 3.23

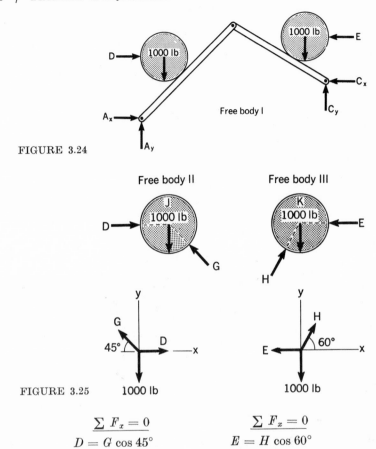

FIGURE 3.24

Free body II   Free body III

FIGURE 3.25   1000 lb   1000 lb

$$\sum F_x = 0 \qquad\qquad \sum F_x = 0$$
$$D = G \cos 45° \qquad\qquad E = H \cos 60°$$
$$D = 1000\ \text{lb} \qquad\qquad E = 578\ \text{lb}$$

Now returning to Fig. 3.24 we note that we still have four unknowns and so we have to examine other free bodies. Accordingly the free-body diagrams of members AB and BC are shown in Fig. 3.26. Here we have two coplanar systems yielding six equations for the six unknowns in the diagrams. Note that the distances from pin A to the contact point of cylinder J and the distance from pin B to the point of contact of cylinder K have been calculated in Fig. 3.27. Since we shall not want forces at pin B, we proceed as follows:

Free body IV:

$$\sum M_B = 0$$
$$(A_y - A_x)7.07 - G(10 - 4.83) = 0$$
$$\therefore A_y - A_x = \frac{1414}{7.07}(5.17) = 1034\ \text{lb} \qquad\qquad \textbf{(a)}$$

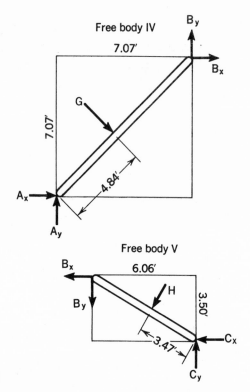

Free body IV

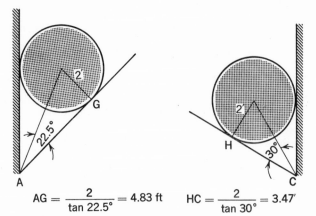

FIGURE 3.26

FIGURE 3.27

$$AG = \frac{2}{\tan 22.5°} = 4.83 \text{ ft}$$ $$HC = \frac{2}{\tan 30°} = 3.47'$$

Free body V:

$$\Sigma M_B = 0$$

$$C_x(3.50) - C_y(6.06) + H(7 - 3.46) = 0$$

$$\therefore 6.06C_y - 3.50C_x = (1155)(3.54) = 4090 \text{ lb} \qquad \text{(b)}$$

Now, going back to free-body diagram I in Fig. 3.24, we get:

Free body I:

$$\Sigma F_y = 0$$

$$A_y + C_y = 2000 \text{ lb} \qquad \text{(c)}$$

$$\Sigma F_x = 0$$

$$A_x - C_x + D - E = 0$$

$$\therefore C_x - A_x = 422 \text{ lb} \qquad \text{(d)}$$

We can now find the desired unknowns by solving Eqs. (a), (b), (c), and (d) simultaneously. We get:

$$A_x = 30 \text{ lb}$$

$$A_y = 1064 \text{ lb} \qquad \text{(e)}$$

$$C_x = 452 \text{ lb}$$

$$C_y = 936 \text{ lb}$$

EXAMPLE 3.5

A derrick is shown in Fig. 3.28 supporting a 1000-lb load. The main beam has a ball socket connection into the ground at d and is held by guy wires.

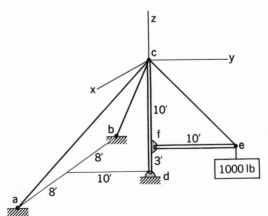

FIGURE 3.28

Neglect the weight of the members and guy wires and ascertain the tensions in the guy wires, *ac*, *bc*, and *ce*.

If we select as a free body both members and the interconnecting guy wire *ce*, we will expose two of the desired unknowns (Fig. 3.29). Note that this is a general three-dimensional force system with only five unknowns. Although all these unknowns can be solved by statical considerations of this free body, you will notice that if we take moments about point *d* we will involve in a vector equation only the desired unknowns T_{bc} and T_{ac}, so all unknown forces need not be computed for this free-body diagram. It is well worth your time to look for such short cuts in situations such as these.

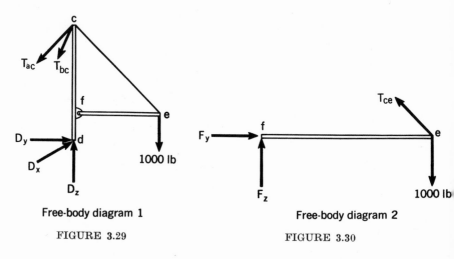

Free-body diagram 1   Free-body diagram 2

FIGURE 3.29   FIGURE 3.30

To determine the unknown tension T_{ce}, we must employ another free-body diagram. Either the vertical or horizontal member will expose this unknown in a manner susceptible to solution. The latter has been selected and is shown in Fig. 3.30. Note that we have here a coplanar force system with three unknowns. Again you can see that by taking moments about point *f* we will involve only the desired unknown.

Returning to Fig. 3.28 we now form a displacement vector $\vec{ca}$ in the direction of T_{ac} as follows:

$$\vec{ca} = (8 - 0)i + (-10 + 0)j + (-13 - 0)k = 8i - 10j - 13k$$

The unit vector in this direction may be formed by dividing by $\sqrt{8^2 + 10^2 + 13^2} = \sqrt{333}$. The vector T_{ac} may then be given as:

$$T_{ac} = T_{ac}\left[\frac{1}{\sqrt{333}}(8i - 10j - 13k)\right] \qquad \text{(a)}$$

Similarly we have for T_{bc}:

$$T_{bc} = T_{bc}\left[\frac{1}{\sqrt{333}}(-8i - 10j - 13k)\right] \qquad \text{(b)}$$

Using the free-body diagram in Fig. 3.29, we now set the sum of moments about point d equal to zero. Thus, employing the above relations, we get:

$$13k \times \left(\frac{T_{ac}}{\sqrt{333}}\right)(8i - 10j - 13k) + 13k \times$$

$$\left(\frac{T_{bc}}{\sqrt{333}}\right)(-8i - 10j - 13k) + (3i + 10j) \times (-1000k) = 0 \qquad \text{(c)}$$

When we introduce $t_1 = T_{ac}/\sqrt{333}$ and $t_2 = T_{bc}/\sqrt{333}$, the preceding equation becomes:

$$[130(t_1 + t_2) - 10,000]i + [104(t_1 - t_2)]j = 0 \qquad \text{(d)}$$

The scalar equations:

$$130(t_1 + t_2) - 10,000 = 0$$

$$104(t_1 - t_2) = 0$$

can now be readily solved to give $t_1 = t_2 = 38.5$. Hence we get $T_{ac} = 38.5\sqrt{333} = 702$ lb and $T_{bc} = 38.5\sqrt{333} = 702$ lb.

Turning finally to the free-body diagram 2 in Fig. 3.30, we see that in summing moments about f the horizontal component of the tension T_{ce} has a zero moment arm. Thus:

$$[(10)(0.707)T_{ce} - (10)(1000)]i = 0$$

Hence:

$$T_{ce} = 1414 \text{ lb}$$

EXAMPLE 3.6

Determine the forces required to support the uniform beam in Fig. 3.31 shown loaded with a couple, a point force, and a parabolic distribution of load. The weight of the beam is given as 100 lb.

Assume that a coplanar force distribution acts on the beam. The free-body diagram is shown in Fig. 3.32. Since there are only three unknown quantities, we may handle the problem by statical consideration of this free body.

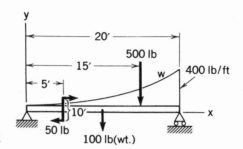

FIGURE 3.31

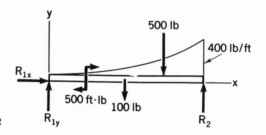

FIGURE 3.32

The equation for the loading curve must be $w = ax^2 + b$, where a and b are to be determined from the loading data and the choice of reference. With an xy reference at the left end, as shown, we then have the conditions:

1. when $x = 0$, $w = 0$
2. when $x = 20$, $w = 400$

To satisfy these conditions b must be zero and a unity; the loading function is thus given as:

$$w = x^2$$

In this problem we shall work directly with the scalar equations. By inspection, we see that in summing forces in the x direction the component $(R_1)_x$ is zero. By summing moments about the left and right ends of the beam, we can then solve the remaining unknowns directly:

$$\underline{\sum M_1 = 0}$$

$$-500 - (10)(100) - (15)(500) - \int_0^{20} x^3 \, dx + 20R_2 = 0$$

Integrating and canceling terms we get:

$$-9000 - \frac{x^4}{4} \bigg]_0^{20} + 20R_2 = 0$$

By inserting limits and solving we get one of the unknowns:

$$R_2 = 2450 \text{ lb}$$

Next:

$\underline{\sum M_2 = 0}$

$$-20(R_1)_y - 500 + (10)(100) + (5)(500) + \int_0^{20} (20 - x)x^2 \, dx = 0$$

Solving for $(R_1)_y$, then, we have:

$$(R_1)_y = 817 \text{ lb}$$

As a check on these computations, we may sum forces in the vertical direction. The result should be close to zero:

$\underline{\sum F_y = 0}$

$$(R_1)_y + R_2 - 100 - 500 - \int_0^{20} x^2 \, dx = 0$$

$$3267 - 600 - \frac{x^3}{3} \bigg]_0^{20} = 0$$

$$\therefore 2667 - 2667 = 0$$

Whenever it is possible to check a solution, we should avail ourselves of the opportunity.

EXAMPLE 3.7

Determine the vertical force F that must be applied to the windlass in Fig. 3.33 to maintain the 100-lb weight. Also estimate the supporting forces from the bearings onto the shaft. The handle DE on which the force is applied is in the indicated xz plane.

To expose the desired unknowns, we shall make a free-body diagram of the windlass without the bearings. To render the problem amenable to purely statical analysis we must make a number of simplifying assumptions. First, we shall neglect friction completely in the bearings. In a properly lubricated system, these forces will be comparatively small. Next, we will assume that the bearings are so narrow that without friction the force distribution from the bearing to the shaft can be considered a coplanar distribution and thus replaceable by a single force normal to the centerline of the shaft. Little error will be incurred if this force is assumed to have a line of action through the

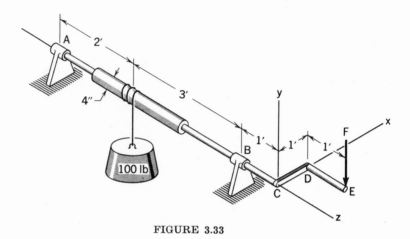

FIGURE 3.33

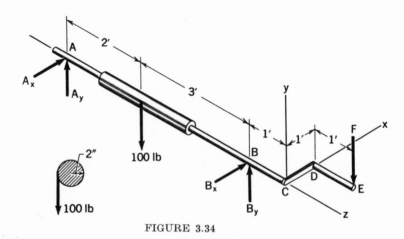

FIGURE 3.34

shaft centerline. Using rectangular components for the bearing forces, we obtain the free-body diagram shown in Fig. 3.34.

Note that we have a general force system in space with five unknowns. Since the sum of the force components in the z direction is already taken to be zero, we will have available five equations of statics. Thus we are just able to handle this problem. It may be clear now why it was so important to simplify the problem at the outset.

Summing the forces we get:

$$(A_x\boldsymbol{i} + A_y\boldsymbol{j}) - 100\boldsymbol{j} - F\boldsymbol{j} + (B_x\boldsymbol{i} + B_y\boldsymbol{j}) = \boldsymbol{0} \qquad \text{(a)}$$

Taking moments about the point B we have:

$$-5k \times (A_x i + A_y j) + (-3k - \tfrac{1}{6}i) \times (-100j) + (2k + i) \times (-Fj) = 0$$

(b)

If we carry out the cross products and rearrange Eq. (b) we get:

$$(5A_y + 2F - 300)i - (5A_x)j + (\tfrac{100}{6} - F)k = 0$$

We have then the following scalar equations from (a) and (b):

$$A_x + B_x = 0 \tag{c}$$

$$A_y + B_y - F - 100 = 0 \tag{d}$$

$$-5A_y - 2F + 300 = 0 \tag{e}$$

$$5A_x = 0 \tag{f}$$

$$\tfrac{100}{6} - F = 0 \tag{g}$$

From equations (f) and (g) we see that A_x and F are 0 and 16.67 lb, respectively. Equation (c) then indicates that B_x is zero. A_y can now be determined from Eq. (e) as 53.3 lb. Finally B_y is evaluated from Eq. (d) as 63.3 lb.

3.6. A Simple Conclusion from Equilibrium

We shall now examine a simple case of equilibrium that occurs quite often and from which simple, useful conclusions may readily be drawn.

Consider a rigid body on which two forces are respectively acting at points a and b. If the body is in equilibrium, the first basic equation of statics, 3.1(a), stipulates that $F_1 = -F_2$; that is, the forces must be equal and opposite. The second fundamental equation of statics, 3.1(b), requires that the forces be collinear so as not to form a nonzero couple. With points a and b given as points of application for the two forces in Fig. 3.35, it is clear that the common line of action for the forces must coincide with the line segment ab.

We often have to deal with pin-connected structural members with loads applied at the pins, as shown in Fig. 3.36. If we neglect friction at the pins and also the weight of the members, we can conclude that only two forces act on each member. These forces, then, must be equal and opposite and must have a line of action collinear with the centerline of the member. The top member in the diagram is a compres-

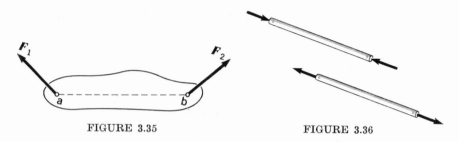

FIGURE 3.35   FIGURE 3.36

sion member, the one below a tensile member. We shall investigate simple structures involving such members in Section 3.8.

EXAMPLE 3.8

Shown in Fig. 3.37 is a device for crushing rocks. A piston D having an 8-in. diameter is activated by a pressure p of 50 psi above that of the atmosphere. Rods AB, BC, and BD can be considered weightless for this problem. What is the horizontal force transmitted at A to the trapped rock shown in the diagram?

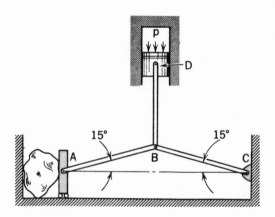

FIGURE 3.37

We have here three two-force members coming together at B. Accordingly, if we isolate pin B as a free body we will have three forces acting on the pin. These forces must be collinear with the centerlines of the respective members as explained earlier (Fig. 3.38).

The force F_D is easily computed by considering the action of the piston. Thus we get:

$$F_D = (50)\left(\frac{\pi 8^2}{4}\right) = 2510\,\text{lb}$$

Summing forces at pin B:

$$\sum F_x = 0$$
$$F_B \cos 15° = F_C \cos 15°$$
$$F_B = F_C$$
$$\sum F_y = 0$$
$$2F_B \sin 15° = 2510 \text{ lb}$$
$$F_B = \frac{2510}{(2)(0.259)} = 4850 \text{ lb}$$

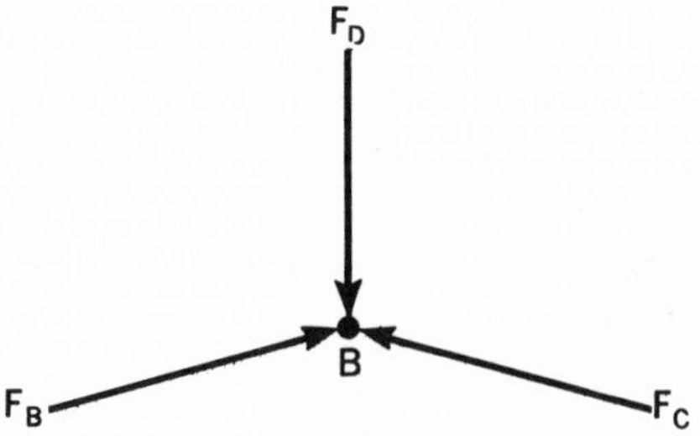

FIGURE 3.38

The force transmitted to the rock in the horizontal direction is then 4850 cos 15° = 4680 lb.

3.7. Static Indeterminacy

Examine the simple beam in Fig. 3.39, with known external loads and weight. If the deformation of the beam is small and the final positions of the external loads differ only slightly from their initial positions, we can assume the beam to be rigid and solve for the supporting forces A, B_x, and B_y, since we have three equations of equilibrium available. Suppose now an additional support is made

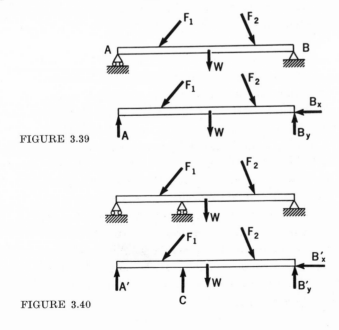

FIGURE 3.39

FIGURE 3.40

available to the beam, as indicated in Fig. 3.40. The beam can still be considered a rigid body, since the applied loads will shift even less because of deformation. Therefore, the equivalent force coming from the ground to counteract the applied loads and weight of the beam must be the same as before. In the first case, in which two supports were given, however, a unique set of values for the forces A, B_x, and B_y gave us the required resistance. In other words, we were able to solve these forces by statics alone without further considerations. In the second case, statics will give the same equivalent supporting force system, but now there are an infinite number of possible combinations of values of the supporting forces that will give us this equivalent system demanded by equilibrium of rigid bodies. To decide the proper combination of supporting forces requires additional computation. Although the deformation properties of the beam were unimportant up to this point, they now become the all-important criterion in apportioning the supporting forces. These problems are termed *statically indeterminate*, in contrast to the statically determinate type in which statics and the rigid-body assumption suffice. It is interesting to note that for a given system of loads and masses two models—the rigid-body model and models taught in other courses involving elastic behavior—may be employed to achieve a desired end. *In summary, we can state that in statically indeterminate problems we must satisfy both the equations of equilibrium for rigid bodies and the equations that stem from deformation considerations in order to determine the unknown forces and couples.*

In the discussion thus far we used a beam as the rigid body and discussed the statical determinacy of the supporting system. It should be clear that the same conclusions apply to any structure that, without the aid of the external constraints, can be taken as a rigid body. If, for such a structure as a free body, there are as many unknown supporting force and couple-moment components as there are equations of equilibrium and if these equations can be solved for these unknowns, then we say that the structure is *externally statically determinate*.

On the other hand, should we desire to know the forces and couples transmitted between internal members of this kind of structure (i.e. one that does not depend on the external constraints for rigidity), we then examine free bodies of these members. When all the unknown force and couple-moment components can be solved by the equations of equilibrium for these free bodies, we then say that the structure is *internally statically determinate*.

There are structures that depend on the external constraints
for rigidity (see the structure
shown in Fig. 3.41). For such
structures, the supporting force
system always depends on both
the internal forces and couples
and the external loads. In this
case we do not distinguish be-

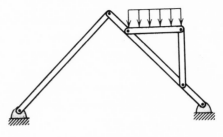

FIGURE 3.41

tween internal and external statical determinacy, since the evalua-
tion of supporting forces and couples will involve free bodies (and hence
forces and couples) for some or all of the internal members of the
structure. For such cases we simply state that the structure is statically
determinate if for all the unknown force and couple-moment components we have enough equations of equilibrium that can be solved
for these quantities.

3.8. Simple Trusses

A *truss* is a system of uniform members (welded, riveted,
or pinned together) that is constructed to support either stationary or
moving loads. In order to design the sizes and shapes of these component
members for a given task, we must be able to compute the forces being
transmitted by each member.

To simplify computations we shall *idealize* the truss. In the general
three-dimensional truss (called a *space truss*), we will imagine that the
members of the truss are interconnected by smooth ball-and-socket
joints, while in the coplanar truss (called a *plane truss*) the members
will be considered to be connected by smooth pins. These models will
give good results, even when some other mode of connection is present,
provided the centerlines of the actual structural members are approximately concurrent at the junctures between members, with pins
and ball joints forming the points of concurrency.

We will further assume that the external loads on the truss are
delivered first to the pins and ball joints. This means that if we neglect
gravity, each member is effectively subjected only to a single force at
each end. Consequently, the forces from a pin or ball joint on each
member must be collinear with the center line of the member, making
each a simple tension or compression member.

An idealized truss is termed *just-rigid* if the removal of any of its members destroys its rigidity. If removing a member does not destroy rigidity, the truss is said to be *over-rigid*. We shall be concerned with just-rigid trusses in this chapter.

The most elemental just-rigid truss is one with three members connected to form a triangle (*ABC* in Fig. 3.42). Other just-rigid space trusses may be built up by adding for each new joint three new members, as is shown in Fig. 3.42. Trusses constructed in this manner are called *simple space trusses*. The *simple plane truss* is constructed by adding for each pin two members always in the plane of the initial triangle, as shown in Fig. 3.43.

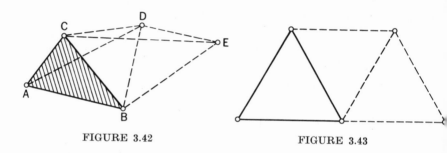

FIGURE 3.42   FIGURE 3.43

Because simple trusses are just-rigid, they are internally statically determinate. Thus, if the supporting force system is statically determinate, we can compute the forces in all the members of such trusses. More specifically, in examining the ball joints of simple space trusses, we can see that in the general three-dimensional case a ball joint with only three unknown forces acting on it from the members can always be found. (It will be the last joint formed.) Each unknown force from a member onto this joint must have a direction collinear with that member and hence has a known direction. There are, then, only three unknown scalars, and they may be solved by statics alone. We can then find another joint with only three unknowns and so carry on the computations until the entire truss has been evaluated. For the simple plane truss, we may proceed with two unknowns at each pin and thus evaluate these forces by statics.

The first operation in ascertaining the forces in the members of a simple truss is generally to compute the supporting forces. We do this by considering the truss as a free body and then employing the equations of equilibrium. The analytical investigation of the truss itself may next be carried out by one of two methods, the difference between the methods being in the choice of the free bodies to be analyzed. The first method uses the hypothetical pins or ball joints as free bodies and is called the *method of joints*. This method has already been discussed briefly in the previous paragraph. In the second method, called the *method of sections*, portions of the entire truss are "cut out" and examined as free bodies.

On the following pages are examples that illustrate both these methods of approach. In the case of the method of joints, you will note that the procedure is self-checking.

In structural work of this kind, we may either neglect the weight of the members, or, if somewhat greater accuracy is desired, it is common practice to take one-half the weight of each member as an external load applied at the pins or ball joints at the ends of the member.

EXAMPLE 3.9

Shown in Fig. 3.44 is a simple plane truss. Two 1000-lb loads are shown acting on pins C and E. We are to determine the force transmitted by each member and to ascertain whether it is a tension or a compression member. Neglect the weight of the members.

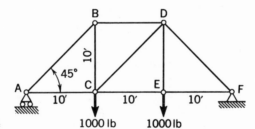

FIGURE 3.44

The first step in a structural analysis is usually to ascertain the supporting forces from the ground, using the truss as a free body. In this simple loading, we see by inspection that there are 1000-lb vertical loads at each support. We shall begin, then, by studying pin A.

Pin A. The forces on pin A are the known 1000-lb supporting force and two unknown forces from the members AB and AC. The direction of these forces is known from the geometry of the truss, but the magnitude and sense must be determined. To help in interpreting the results, put the forces in the same position as the corresponding members in the space diagram (Fig. 3.45). That is, avoid the force diagram in Fig. 3.46, which is equivalent to the one in Fig. 3.45 but which may lead to errors in interpretation. There

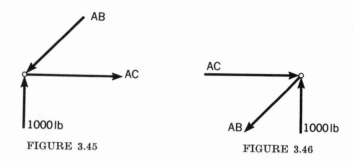

FIGURE 3.45   FIGURE 3.46

are two unknowns for the concurrent coplanar force system in Fig. 3.45 and thus if we use the scalar equations of equilibrium we may evaluate AB and AC:

$\underline{\sum F_y = 0}$:

$$-0.707AB + 1000 = 0$$
$$AB = 1414 \text{ lb}$$

$\underline{\sum F_x = 0}$:

$$AC - 0.707AB = 0$$
$$AC = 1000 \text{ lb}$$

Observing pin A in Fig. 3.45 we see that a force of 1414 lb is being exerted on the pin from member AB in such a manner as to "push" the pin. Thus member AB is a compression member since such a deformation is needed in the beam to deliver this action. Since member AC is "pulling" on the pin A with a force of 1000 lb, it is a tension member. (We can now appreciate why the first free-body diagram is to be preferred.) As a help in keeping the compressive or tensile nature of these members clear, we can use the diagrammatic aid in Fig. 3.47 when the computations for pin A are made and fully diagnosed.

If we next examine pin C, it becomes clear that since there are three unknowns involved for this pin we cannot solve the forces by equilibrium equations. However pin B can be handled, and once BC is known pin C can be examined.

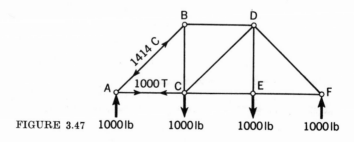

FIGURE 3.47   1000 lb   1000 lb   1000 lb   1000 lb

Pin B. Summing forces on pin B (Fig. 3.48) we get:

<u>$\Sigma\, F_y = 0$:</u>

$$(1414)(0.707) + BC = 0$$
$$BC = -1000\ \text{lb}$$

<u>$\Sigma\, F_x = 0$:</u>

$$(1414)(0.707) + BD = 0$$
$$BD = -1000\ \text{lb}$$

Here we have incurred two minus quantities, indicating that we have made an incorrect choice of sense. The correct free body of pin B is then given in Fig. 3.49. Note that the same orientation relative to the space diagram has been maintained. The member BD is a compression member, while BC is a tension member.

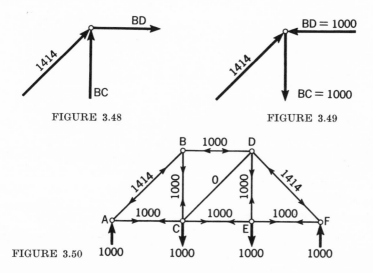

FIGURE 3.48   FIGURE 3.49

FIGURE 3.50

We can proceed in this manner from joint to joint. At the last joint a check is possible since all the forces will have been computed for this pin. The final solution is shown in Fig. 3.50.

EXAMPLE 3.10

If the force transmitted by a single member is desired, we can employ a more suitable free body for using the equations of equilibrium. Suppose in the previous example we wish to know the force in member CE only. To avoid the laborious joint-by-joint procedure, we employ a portion of the truss as shown in Fig. 3.51. Notice that the forces from the other part of the

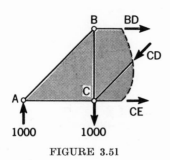

FIGURE 3.51

truss acting on this part through the cut members have been included, and in this way the desired force has been exposed. The sense of these exposed forces is not known but we do know the directions, as explained in our earlier discussions. Using the equations of equilibrium and taking advantage of the fact that the lines of action of some of the exposed unknown forces are concurrent at certain joints, we may readily solve for the unknowns if they number three or less. To determine CE we take moments about a point corresponding to joint D. This will give an equation involving only CE as the unknown:

$$\sum M_D = 0:$$
$$(1000)(20) - (1000)(10) - 10CE = 0$$
$$CE = 1000 \text{ lb}$$

By observing the force diagram in Fig. 3.51, we can clearly see that the member is a tension member.

If we desire BD also, we can take moments about point C. However, this time BD comes out negative, indicating that we have made an incorrect choice of sense. We must change this and then determine whether the bar is tensile or compressive, which we can readily do by inspection.

It may be that a suitable section with sufficient unknowns for a solution cannot be made, and we may have to take several sections before we can expose the desired force in a free body with enough simultaneous equations to effect a solution. These problems are no different from the ones we studied earlier in this chapter, where several free-body diagrams were needed to generate a complete set of equations containing the unkown quantity.

EXAMPLE 3.11

Shown in Fig. 3.52 is a plane truss for which the force in member AB only is desired. The supporting forces have been determined and are shown in the diagram.

In Fig. 3.53 we have shown a cut section of the truss exposing force AB. (This gives the same force diagram as that which results from the free-body diagram of pin A.) We have here three unknown forces for which only two equations of equilibrium are available. We must use an additional free body.

Thus, in Fig. 3.54 we have shown a second cut section. Note that by taking moments about joint B we can solve for AC, and with this information

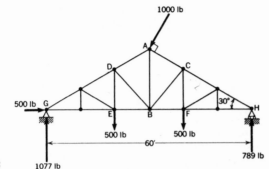

FIGURE 3.52

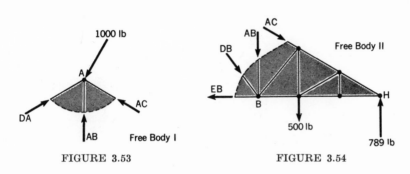

FIGURE 3.53   FIGURE 3.54

we can return to the first free body to get the desired unknown AB. Accordingly, we have for free body II:

$$\underline{\sum M_B = 0}:$$
$$(10)(500) - (30)(789) + (AC)(\sin 30°)(30) = 0$$

Note we have transmitted AC to joint H in evaluating its moment contribution. Solving for AC we get

$$AC = 1245 \text{ lb}$$

Summing forces for free body I we have:

$$\underline{\sum F_x = 0}:$$
$$DA \cos 30° - AC \cos 30° - 500 = 0$$
$$\therefore DA = 1821 \text{ lb}$$

$$\underline{\sum F_y = 0}:$$
$$DA \sin 30° + AC \sin 30° + AB - 866 = 0$$
$$\therefore AB = -672 \text{ lb}$$

We see that member AB is a tension member rather than a compression member as was our initial guess in drawing the free-body diagrams.

EXAMPLE 3.12

Ascertain the forces transmitted by each member of the three-dimensional truss (Fig. 3.55).

We can readily find the supporting forces for this simple structure by considering the whole structure as a free body and by making use of the symmetry of the loading and geometry (Fig. 3.56).

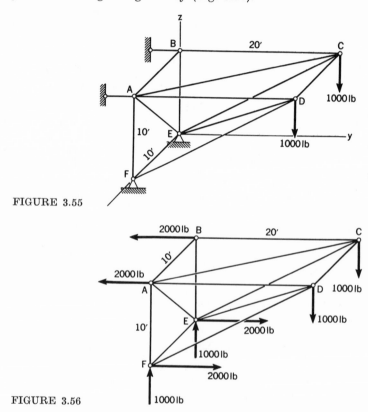

FIGURE 3.55

FIGURE 3.56

Joint F. It is clear, on an inspection of joint F, that the force FE must be zero, since all other forces are in a plane at right angles to it. These other forces are shown in Fig. 3.57. Summing forces in the x and y directions, we get:

$$\underline{\sum F_y = 0:}$$

$$FD \frac{20}{\sqrt{20^2 + 10^2}} = 2000$$

$$FD = 2240 \text{ lb}$$

$$\Sigma \, F_z = 0:$$

$$-AF + 1000 - 2240 \, \frac{10}{\sqrt{500}} = 0$$

$$AF = 1000 - 1000 = 0$$

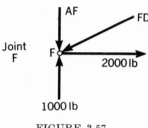

Joint F

Joint B. Going to joint B, we see that $AB = 0$ and $BE = 0$, since there are no other force components on pin B in the directions of these members. Finally, $BC = 2000 \,\text{lb}$ tension.

FIGURE 3.57

Joint A. Let us next consider joint A (Fig. 3.58). We will first determine the unit vectors $\hat{\boldsymbol{\rho}}_{AC}$ and $\hat{\boldsymbol{\rho}}_{AE}$ so that we can express forces $\overrightarrow{AC}$ and $\overrightarrow{AE}$ vectorially. Thus:

$$\boldsymbol{\rho}_{AC} = \boldsymbol{r}_C - \boldsymbol{r}_A = (20\boldsymbol{j} + 10\boldsymbol{k}) - (10\boldsymbol{i} + 10\boldsymbol{k})$$

$$= -10\boldsymbol{i} + 20\boldsymbol{j}$$

$$\therefore \hat{\boldsymbol{\rho}}_{AC} = \frac{1}{\sqrt{500}} (-10\boldsymbol{i} + 20\boldsymbol{j})$$

$$\boldsymbol{\rho}_{AE} = \boldsymbol{r}_E - \boldsymbol{r}_A = \boldsymbol{0} - (10\boldsymbol{i} + 10\boldsymbol{k}) = -10\boldsymbol{i} - 10\boldsymbol{k}$$

$$\therefore \hat{\boldsymbol{\rho}}_{AE} = \frac{1}{\sqrt{200}} (-10\boldsymbol{i} - 10\boldsymbol{k})$$

Accordingly we have:

$$\overrightarrow{AC} = \frac{AC}{\sqrt{500}} (-10\boldsymbol{i} + 20\boldsymbol{j})$$

$$\overrightarrow{AE} = \frac{AE}{\sqrt{200}} (-10\boldsymbol{i} - 10\boldsymbol{k})$$

Summing forces, we have:

$$-2000\boldsymbol{j} + AD\boldsymbol{j} + AC(-0.446\boldsymbol{i} + 0.892\boldsymbol{j}) + AE(-0.707\boldsymbol{i} - 0.707\boldsymbol{k}) = \boldsymbol{0}$$

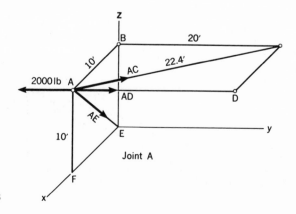

FIGURE 3.58

Hence:

$$0.892AC + AD = 2000 \tag{a}$$

$$-0.446AC - 0.707AE = 0 \tag{b}$$

$$-0.707AE = 0 \tag{c}$$

We see that $AE = AC = 0$ and $AD = 2000$ lb tension.

Joint D. We now consider joint D (Fig. 3.59). The unit vectors to be used with forces $\overrightarrow{FD}$ and $\overrightarrow{ED}$ are determined as follows:

$$\boldsymbol{\rho}_{ED} = \boldsymbol{r}_E - \boldsymbol{r}_D = \boldsymbol{0} - (10\boldsymbol{i} + 20\boldsymbol{j} + 10\boldsymbol{k})$$

$$= -10\boldsymbol{i} - 20\boldsymbol{j} - 10\boldsymbol{k}$$

$$\therefore \hat{\boldsymbol{\rho}}_{ED} = \frac{1}{\sqrt{600}}(-10\boldsymbol{i} - 20\boldsymbol{j} - 10\boldsymbol{k})$$

$$\boldsymbol{\rho}_{FD} = \boldsymbol{r}_D - \boldsymbol{r}_F = (10\boldsymbol{i} + 20\boldsymbol{j} + 10\boldsymbol{k}) - (10\boldsymbol{i})$$

$$= 20\boldsymbol{j} + 10\boldsymbol{k}$$

$$\therefore \hat{\boldsymbol{\rho}}_{FD} = \frac{1}{\sqrt{500}}(20\boldsymbol{j} + 10\boldsymbol{k})$$

Accordingly:

$$\overrightarrow{ED} = \frac{ED}{\sqrt{600}}(-10\boldsymbol{i} - 20\boldsymbol{j} - 10\boldsymbol{k})$$

$$\overrightarrow{FD} = \frac{FD}{\sqrt{600}}(20\boldsymbol{j} + 10\boldsymbol{k})$$

Hence, summing forces, we get:

$$-2000\boldsymbol{j} - 1000\boldsymbol{k} - DC\boldsymbol{i} + 2240(0.893\boldsymbol{j} + 0.446\boldsymbol{k})$$

$$+ ED(-0.406\boldsymbol{i} - 0.816\boldsymbol{j} - 0.406\boldsymbol{k}) = \boldsymbol{0} \tag{d}$$

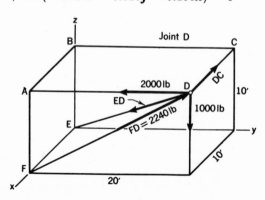

FIGURE 3.59

Thus:

$$-2000 + 2000 - 0.816ED = 0 \qquad \text{(e)}$$

$$DC + 0.406ED = 0 \qquad \text{(f)}$$

$$-1000 + 1000 - 0.406ED = 0 \qquad \text{(g)}$$

We see here that $ED = 0$ and $DC = 0$.

Joint E. The only nonzero forces on joint E are the supporting forces and $\overline{CE}$, as shown in Fig. 3.60. We may solve for CE directly and get 2240 lb compression.

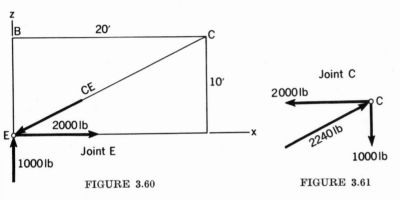

FIGURE 3.60   FIGURE 3.61

Joint C. As a check on our problem, we can examine joint C. The only nonzero forces are shown on the joint (Fig. 3.61). The reader may readily verify that the solution checks.

The results for the truss are then:

$AF = 0,$	$ED = 0,$
$FE = 0,$	$DC = 0,$
$FD = 2240$ lb compression,	$AD = 2000$ lb tension
$AB = 0,$	$AE = 0,$
$AC = 0,$	$BE = 0,$
$BC = 2000$ lb tension,	$CE = 2240$ lb compression

3.9. Friction

Friction may be defined as the force distribution at a surface of contact between two bodies that prevents or impedes any possible sliding motion between the two bodies. This force distribution is tangent to the contact surface and has, for the body under consideration, a direction at every point in the contact surface that is in opposition to the possible or existing relative slipping motion at that point.

Coulomb friction is that friction which occurs between bodies having dry contact surfaces and is not to be confused with the action of one body on another separated by a film such as oil. The latter problems are termed lubrication problems and are examined in courses on fluid mechanics. Coulomb, or dry, friction is a complicated phenomenon and actually not much is known about its true nature.* It is believed that the major cause of this friction is the microscopic roughness of the surfaces of contact. Interlocking microscopic protuberances oppose the relative motion between the surfaces. When sliding is present between the surfaces, some of these protuberances either are sheared off or are melted by high local temperatures. This is the reason for the high rate of "wear" for dry-body contact and indicates why it is desirable to separate the surfaces by a film of fluid.

We have previously employed the terms "smooth" and "rough" surfaces of contact. The former means that only a force normal to the contact surface can be transmitted between the bodies, while in the latter case a tangential force (i.e., a friction force) can also be present. We shall now present an approximate relation which may, under certain conditions, relate the normal and frictional forces at a point of contact between two nonlubricated rigid bodies.

Everybody has gone through the experience of sliding furniture along a floor. We exert a continuously increasing force which is completely resisted by friction until the object begins to move—usually with a lurch. The lurch occurs because once the object begins to move

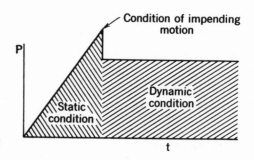

FIGURE 3.62

---

*For more complete discussion of friction, see F. P. Bowden anc D. Tabor, *The Friction and Lubrication of Solids* (New York: Oxford University Press, 1950).

there is a decrease in frictional effect from the maximum effect attained under static conditions. An idealized plot of this action as a function of time is shown in Fig. 3.62, which shows the drop from the highest or limiting frictional effect to a frictional effect that is constant with time, i.e., independent of the velocity of the object and with a magnitude less than the maximum static value. The condition corresponding to this maximum is termed *condition of impending motion.*

Consider a plane surface of contact between two bodies. At the *condition of impending translational relative motion** and *once this motion has begun,* it is possible to relate the frictional and normal components between the bodies as a result of studies carried out by Coulomb in 1781. For such situations, he reported in effect that:

1. The total amount of friction that can be developed is independent of the magnitude of the area of contact.

2. The total frictional force that can be developed is proportional to the normal force transmitted across the surface of contact.

3. For low velocities, the total amount of friction that can be developed is practically independent of velocity. However, it is less than the frictional force corresponding to impending motion.

Conclusions 1 and 3 may come as a surprise to most of you and be contrary to your "intuition." Nevertheless, they are accurate enough statements for many engineering applications. More precise studies of friction, as was pointed out earlier, are complicated and involved. We may express Conclusion 2 mathematically:

$$f \propto N, \qquad \therefore f = \mu N \qquad\qquad \textbf{3.14}$$

where μ is called the *coefficient of friction.*

It must be strongly cautioned that this relation is valid only at conditions of impending motion or while the body is in motion. A careful rereading of Coulomb's laws confirms this. Since limiting static friction exceeds dynamic friction, we differentiate between coefficients of friction for those conditions. Thus we have coefficients of static friction and coefficients of dynamic friction, μ_s and μ_d, respectively.

---

*This means that all the points on any one of the two contact surfaces must have an actual or impending relative motion that is the same for all points on the surface with respect both to magnitude and direction. In other words the contact surfaces cannot be on the verge of rotating relative to each other or actually be rotating relative to each other. We shall examine such cases later in this section.

The accompanying table is a small list of coefficients. The corresponding coefficients of friction for dynamic conditions are about 25 per cent less.

STATIC COEFFICIENTS OF FRICTION*

Steel on cast iron	0.40
Copper on steel	0.36
Hard steel on hard steel	0.42
Mild steel on mild steel	0.57
Rope on wood	0.70
Wood on wood	0.20–0.75

We shall now examine a number of statics problems that involve plane surfaces of contact between bodies for which there is the possibility of relative translational motion. The contact surfaces may at times be small enough to be considered point contacts or they may at other times be considered finite. In this regard it may be well to point out that the normal force distribution in either case is a parallel force distribution replaceable by a single force N. However, because we generally do not know the *distribution* of the normal force over a finite contact surface, we do not know the *line of action* for N. This restricts us in such cases to using only the summation equations of equilibrium. (Clearly, using a moment equation would serve only to bring in an unknown distance.) This, of course, is not true for point contact between bodies, since the line of action of N must pass through the point of contact.

Two common types of statics problems involve dry friction. In the first type, we know that motion is impending, or has been established and is uniform, and we desire information about certain forces that are present. We can then relate normal and friction forces at surfaces of contact where there is impending or actual relative motion using Coulomb's laws and, employing f_i for other friction forces, proceed by methods of statics. However, the *proper direction must be given to all friction forces*. That is, they must *oppose* possible, impending, or actual relative motion at the contact surfaces. In the second type of friction problem, external loads on a body are given and we desire to deter-

---

*F. P. Bowden and D. Tabor, *The Friction and Lubrication of Solids* (New York: Oxford University Press, 1950).

mine whether the friction forces present are sufficient to maintain equilibrium. One way to attack this latter type of problem is to assume that impending motion exists in the various directions possible and to solve for the external forces required for such conditions. By comparing the actual external forces present with those required for the various impending motions, we can then deduce whether the body can be restrained by frictional forces from sliding. The following problems illustrate these cases.

EXAMPLE 3.13

In Fig. 3.63 (a) is shown an automobile on a roadway inclined at an angle θ with the horizontal. If the coefficients of static and dynamic friction between the tires and the road are taken as 0.6 and 0.5, respectively, what is the maximum inclination $\theta_{\max}$ that the car can climb? It has rear wheel drive

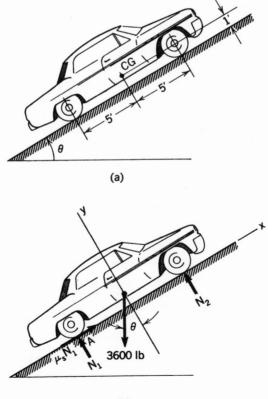

(a)

(b)

FIGURE 3.63

and has total weight of 3600 lb when loaded. The center of gravity for this loaded condition has been shown in the diagram.

Let us assume that the drive wheels do not "spin,"—that is, there is zero relative motion between the tire surface and the road surface at the point of contact. Then clearly the maximum friction force possible is μ_s times the normal force at this contact surface. This has been indicated in Fig. 3.63 (b).

We can consider this to be a coplanar problem with three unknowns, namely N_1, N_2, and θ_{max}. Accordingly, since the friction force is restricted to a point, we have three equations of equilibrium available. Using the reference xy shown in the diagram, we have:

$\underline{\sum F_x = 0}$:

$$0.6N_1 - 3600 \sin \theta_{max} = 0 \tag{a}$$

$\underline{\sum F_y = 0}$:

$$N_1 + N_2 - 3600 \cos \theta_{max} = 0 \tag{b}$$

$\underline{\sum M_A = 0}$:

$$10N_2 + (3600 \cos \theta_{max})(5) + (3600 \sin \theta_{max})(1) = 0 \tag{c}$$

To solve for θ_{max} we eliminate N_1 from Eqs. (a) and (b), getting as a result the equation:

$$N_2 = 3600 \cos \theta_{max} - 6000 \sin \theta_{max} \tag{d}$$

Now, eliminating N_2 from Eqs. (c) and (d), we get:

$$18,000 \cos \theta_{max} - 56,400 \sin \theta_{max} = 0 \tag{e}$$

Hence:

$$\tan \theta_{max} = 0.319$$
$$\therefore \theta_{max} = 17.7° \tag{f}$$

If the drive wheels were caused to spin, we would have to use μ_d in place of μ_s for this problem. We would then arrive at a smaller θ_{max}, which for this problem would be 14.7°.

EXAMPLE 3.14

Using the data of Example 3.13, compute the torque needed by the drive wheels to move the car at a uniform speed up an incline where $\theta = 15°$. Also, assume the brakes have "locked" while the car is in a parked position

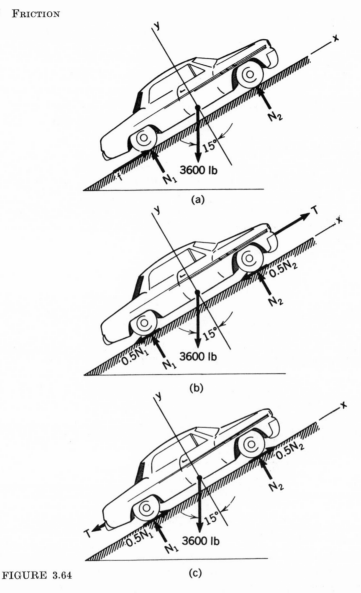

FIGURE 3.64

(a)

(b)

(c)

on the incline. What force is then needed to tow the car either up the incline or down the incline with the brakes in this condition? The diameter of the tire is 25 inches.

A free-body diagram for the first part of the problem is shown in Fig. 3.64 (a). Note that the friction force f will now be determined by Newton's law and not by Coulomb's law since we do not have impending slippage between the wheel and the road for this case. Accordingly, we have for f:

$$\underline{\sum F_x = 0:}$$

$$f - 3600 \sin 15° = 0$$

$$\therefore f = 930 \text{ lb}$$

The torque needed is then:

$$\text{torque} = (f)(r) = (930)\left(\frac{\frac{25}{2}}{12}\right) = 1162 \text{ ft-lb}$$

For the second part of the problem, we have shown the required free body in Fig. 3.64 (b). Note that we have used Coulomb's law for the friction force with the dynamic friction coefficient μ_d. We now write the equations of equilibrium for this free body:

$$\underline{\sum F_x = 0:}$$

$$T - 0.5(N_1 + N_2) - 1230 = 0 \tag{a}$$

$$\underline{\sum F_y = 0:}$$

$$(N_1 + N_2) - (3600)(0.940) = 0 \tag{b}$$

Solving for $N_1 + N_2$ from Eq. (b), and substituting into Eq. (a), we can now solve for T. Hence:

$$T = (0.5)(3600)(0.940) + 1230 = 2920 \text{ lb} \tag{c}$$

For towing the car down the incline we must reverse the direction of the friction force as shown in Fig. 3.64 (c). Solving for T as in the previous example, we get:

$$T = (0.5)(3600)(0.940) - 1230 = 460 \text{ lb} \tag{d}$$

EXAMPLE 3.15

A 15-ft ladder weighing 50 lb is placed in a position where its inclination to a vertical wall is at an angle of 30° (Fig. 3.65). A 200-lb man is to climb the ladder. At what position will he induce slipping? The coefficient of static friction is known to be 0.20 for contacts at the wall and floor.

For impending motion, both surfaces of contact must have developed maximum friction in this problem. We make use of this in the free-body diagram of the ladder in Fig. 3.66. Note that friction forces have been replaced by μ_s times the normal force at each contact surface, and the proper direction of the friction has been shown to prevent motion. The surfaces of contact may be assumed to be infinitesimal; thus the lines of action of the

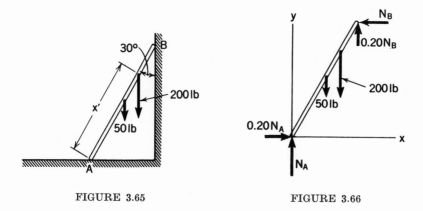

FIGURE 3.65   FIGURE 3.66

normal and friction forces are known for this problem. This may be considered a coplanar problem and since there are three unknown quantities
it is completely solvable. We use the basic vector equations of equilibrium.
First:

$$\mu_s N_A \boldsymbol{i} - N_B \boldsymbol{i} + N_A \boldsymbol{j} - 50\boldsymbol{j} - 200\boldsymbol{j} + \mu_s N_B \boldsymbol{j} = \boldsymbol{0}$$

This means that:

$$\mu_s N_A - N_B = 0 \tag{a}$$

$$N_A - 50 - 200 + \mu_s N_B = 0 \tag{b}$$

giving us the scalar equations for evaluating N_A and N_B.

Taking moments about position A and using the simplest position
vectors to the lines of action of the forces, we get:

$$(0.866)(15)\boldsymbol{j} \times (-N_B)\boldsymbol{i} + (0.5)(15)\boldsymbol{i} \times (0.20N_B)\boldsymbol{j} + (0.5)(x')\boldsymbol{i} \times (-200)\boldsymbol{j}$$
$$+ (0.5)(15/2)\boldsymbol{i} \times (-50)\boldsymbol{j} = \boldsymbol{0}$$

Carrying out the cross products, we have:

$$13N_B \boldsymbol{k} + 1.5N_B \boldsymbol{k} - 100x'\boldsymbol{k} - 187\boldsymbol{k} = \boldsymbol{0}$$

The resulting scalar equation then becomes:

$$14.5N_B - 100x' - 187 = 0 \tag{c}$$

We may readily solve Eqs. (a), (b), and (c) simultaneously. The results
are:

$$N_A = 240 \text{ lb}, \qquad N_B = 48 \text{ lb}, \qquad x' = 5.09 \text{ ft}$$

EXAMPLE 3.16

In Fig. 3.67, a strongbox weighing 150 lb rests on a floor. The static coefficient for the contact surface is 0.20. What is the highest central position for a horizontal load P that permits it to just move the box without tipping it?

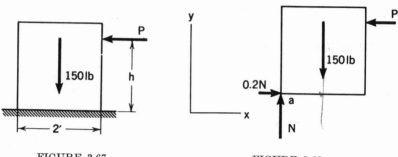

FIGURE 3.67   FIGURE 3.68

The free-body diagram for the condition that gives the limiting position of P is shown in Fig. 3.68. The condition of impending tipping requires that the friction and the normal force be concentrated at the left edge of the box, while the condition of impending slippage is accounted for by using Coulomb's law there.

We may again consider this to be a coplanar system of forces at the midplane of the strongbox. Summing forces, we get the vector equation:

$$0.2N\boldsymbol{i} + N\boldsymbol{j} - 150\boldsymbol{j} - P\boldsymbol{i} = \mathbf{0}$$

Hence:

$$P = 30 \text{ lb}, \qquad N = 150 \text{ lb}$$

Now, taking moments about the corner a, we have:

$$h\boldsymbol{j} \times (-30\boldsymbol{i}) + \boldsymbol{i} \times (-150\boldsymbol{j}) = \mathbf{0}$$

Therefore:

$$30h\boldsymbol{k} - 150\boldsymbol{k} = \mathbf{0}$$

$$h = 5 \text{ ft}$$

Thus the height of the load must be less than 5 ft.

EXAMPLE 3.17

The coefficient of static friction for all contact surfaces in Fig. 3.69 is 0.2. Does the 50-lb load move the block A up, hold it in equilibrium, or is it too small to prevent A from coming down and B from moving out? The 50-lb force is exerted at the midplane of the blocks so that we may consider this a coplanar problem.

We can compute a force P in place of the 50-lb force to cause impending motion of block B to the left and a force P for impending motion to the right. In this way, we can ascertain the action that a 50-lb force will cause.

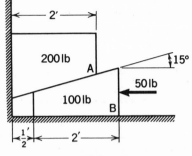

FIGURE 3.69

The free-body diagrams for impending motion to the left have been shown in Fig. 3.70, which contains the unknown force P mentioned above. We need not be concerned about the correct location of the centers of gravity of the blocks, since we shall only add forces in the analysis. This results from the fact that we do not know the line of action of the forces at the contact surfaces and therefore cannot take moments. Summing forces on block A, we get:

$$N_2 \cos 15°\boldsymbol{j} - N_2 \sin 15°\boldsymbol{i} - 0.2N_2\boldsymbol{j} - 200\boldsymbol{j}$$
$$+ N_1\boldsymbol{i} - 0.2N_2 \cos 15°\boldsymbol{i} - 0.2N_2 \sin 15°\boldsymbol{j} = \boldsymbol{0}$$

The scalar equations are:

$$0.966N_2 - 0.2N_1 - 200 - 0.0518N_2 = 0$$
$$N_1 - 0.259N_2 - 0.193N_2 = 0$$

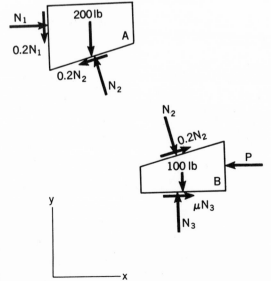

FIGURE 3.70

Solving simultaneously, we get:

$$N_2 = 243 \text{ lb}, \qquad N_1 = 110 \text{ lb}$$

For the free-body diagram of B we have, on summing forces:

$$-N_2 \cos 15° \boldsymbol{j} + N_2 \sin 15° \boldsymbol{i} - P\boldsymbol{i} + 0.2N_3\boldsymbol{i} + N_3\boldsymbol{j}$$
$$- 100\boldsymbol{j} + 0.2N_2 \cos 15° \boldsymbol{i} + 0.2N_2 \sin 15° \boldsymbol{j} = \boldsymbol{0}$$

This yields the following scalar equations:

$$-P + 62.9 + 0.2N_3 + 46.9 = 0$$
$$-235 + N_3 - 100 + 12.6 = 0$$

Solving simultaneously, we have:

$$P = 174 \text{ lb}$$

It becomes clear that a force of 50 lb, as stipulated for this problem, will not induce a motion on block B to the left, so further computation is necessary.

Next let us see what force P is required to move the block to the right. The frictional forces in Fig. 3.70 are all reversed, and the vector equation of equilibrium for block A becomes:

$$N_2 \cos 15° \boldsymbol{j} - N_2 \sin 15° \boldsymbol{i} + 0.2N_1\boldsymbol{j} - 200\boldsymbol{j}$$
$$+ N_1\boldsymbol{i} + 0.2N_2 \cos 15° \boldsymbol{i} + 0.2N_2 \sin 15° \boldsymbol{j} = \boldsymbol{0}$$

The scalar equations are:

$$0.966N_2 + 0.2N_1 - 200 + 0.0518N_2 = 0$$
$$N_1 - 0.259N_2 + 0.193N_2 = 0$$

Solving simultaneously, we get:

$$N_2 = 194.1 \text{ lb}, \qquad N_1 = 12.80 \text{ lb}$$

For free body B we have, on summing forces:

$$-N_2 \cos 15° \boldsymbol{j} + N_2 \sin 15° \boldsymbol{i} - P\boldsymbol{i} - 0.2N_3\boldsymbol{i} + N_3\boldsymbol{j}$$
$$- 100\boldsymbol{j} - 0.2N_2 \cos 15° \boldsymbol{i} - 0.2N_2 \sin 15° \boldsymbol{j} = \boldsymbol{0}$$

The following are the scalar equations:

$$-P - 0.2N_3 + 50.3 - 37.5 = 0$$
$$-100 + N_3 - 187.2 - 10.05 = 0$$

Solving, we get $P = -46.6$ lb. This indicates that we would have to pull to the right to get block B to move in this direction. Thus we may conclude from this study that the blocks are in equilibrium.

---

*EXAMPLE 3.18

Shown in Fig. 3.71 is the problem of mountain climbers trying to move a 200-lb weight W around a chasm in a glacial region. The weight is held by a cable CD fixed at C and fastened to the weight at D. A force F is exerted by the climbers on the other cable. For the position shown in the diagram, what force F must be exerted to start the weight moving from its present position? The coefficient of static friction for this case is to be taken as 0.1. Both cables are to be considered parallel to plane ABE.

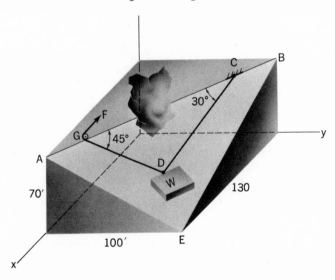

FIGURE 3.71

We have shown a free-body diagram of the load in Fig. 3.72. It is important to realize that the impending motion must be such as not to extend cable CD and so the friction force must be at right angles to the direction of CD in the plane AEB.

By summing forces in the direction normal to the plane, we shall be able to solve for N and consequently get the magnitude of the friction force. For this we shall need the unit vector $\hat{n}$ normal to the plane ABE. To get it, note that the area vector $\overrightarrow{ABE}$ can be expressed in the following manner:

$$\overrightarrow{ABE} = \tfrac{1}{2}(70)(100)\boldsymbol{i} + \tfrac{1}{2}(130)(70)\boldsymbol{j} + \tfrac{1}{2}(130)(100)\boldsymbol{k}$$
$$= 3500\boldsymbol{i} + 4550\boldsymbol{j} + 6500\boldsymbol{k}$$

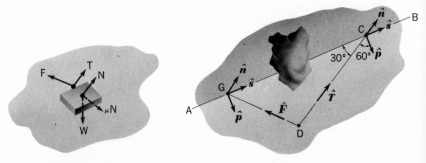

FIGURE 3.72   FIGURE 3.73

Accordingly, we have for $\hat{\boldsymbol{n}}$:

$$\hat{\boldsymbol{n}} = \frac{\overrightarrow{ABE}}{ABE} = \frac{3500}{ABE}\boldsymbol{i} + \frac{4550}{ABE}\boldsymbol{j} + \frac{6500}{ABE}\boldsymbol{k}$$

$$\therefore \hat{\boldsymbol{n}} = 0.404\boldsymbol{i} + 0.525\boldsymbol{j} + 0.750\boldsymbol{k} \tag{a}$$

We may now state on summing forces in the direction $\hat{\boldsymbol{n}}$:

$$N = 200\boldsymbol{k} \cdot \hat{\boldsymbol{n}} = 150 \,\text{lb} \tag{b}$$

We shall also need the directions of $\boldsymbol{T}$ (the force exerted by cable DC), $\boldsymbol{F}$ (the force exerted by cable GD), and the friction force. For this purpose, we have shown at C and G [see Fig. 3.73] a set of orthogonal unit vectors $\hat{\boldsymbol{n}}, \hat{\boldsymbol{s}}$, and $\hat{\boldsymbol{p}}$. We have already computed $\hat{\boldsymbol{n}}$. The vector $\hat{\boldsymbol{s}}$ lying along the line AB is computed as follows:

$$\hat{\boldsymbol{s}} = \frac{(\boldsymbol{r}_B - \boldsymbol{r}_A)}{|\boldsymbol{r}_B - \boldsymbol{r}_A|} = \frac{100}{\sqrt{100^2 + 130^2}}\boldsymbol{j} + \frac{-130}{\sqrt{100^2 + 130^2}}\boldsymbol{i}$$

$$= -0.793\boldsymbol{i} + 0.610\boldsymbol{j} \tag{c}$$

The third vector $\hat{\boldsymbol{p}}$ lying in the plane AEB is computed in the following manner:

$$\hat{\boldsymbol{p}} = \hat{\boldsymbol{s}} \times \hat{\boldsymbol{n}} = [-0.793\boldsymbol{i} + 0.610\boldsymbol{j}]$$
$$\times [0.404\boldsymbol{i} + 0.525\boldsymbol{j} + 0.750\boldsymbol{k}]$$

$$\therefore \hat{\boldsymbol{p}} = 0.458\boldsymbol{i} + 0.595\boldsymbol{j} - 0.662\boldsymbol{k} \tag{d}$$

The unit vector $\hat{\boldsymbol{T}}$ along DC may now easily be computed. Thus:

$$\hat{\boldsymbol{T}} = (\cos 30°)\hat{\boldsymbol{s}} - (\sin 30°)\hat{\boldsymbol{p}} \tag{e}$$

Substituting for $\hat{\boldsymbol{s}}$ and $\hat{\boldsymbol{p}}$ using Eqs. (c) and (d), we get:

$$\hat{\boldsymbol{T}} = -0.917\boldsymbol{i} + 0.231\boldsymbol{j} + 0.331\boldsymbol{k} \tag{f}$$

By a similar procedure we may solve for the unit vector $\hat{F}$ along DG as follows:

$$\hat{F} = 0.237i - 0.853j + 0.468k \tag{g}$$

To get the direction of the friction force, we note that it is at right angles to both $\hat{T}$ and $\hat{n}$. Accordingly, we have for $\hat{f}$, the unit vector in the direction of the friction force:

$$\hat{f} = \hat{n} \times \hat{T} = -0.821j + 0.575k \tag{h}$$

We now employ the equilibrium equation in the direction of $\hat{f}$. Thus, we get:

$$-0.1N + (F\hat{F}) \cdot \hat{f} + (-200k) \cdot \hat{f} = 0 \tag{i}$$

Making the appropriate substitutions from previous results, we get:

$$F = 134 \text{ lb}$$

In the examples undertaken, the nature of the relative impending or actual motion between the surfaces of contact was quite simple. We shall now examine more general types of contacts between bodies. In Example 3.19 we have a plane contact surface but with varying direction of impending or slipping motion for the area elements. In Example 3.20 we have, in addition to the aforementioned complication, the added complexity of a noncoplanar contact surface. Finally, Example 3.21 is a friction problem calling for vector techniques involving a body with significant spatial dimensions—in contrast to the body of Example 3.18, which was taken as a particle in the analysis. In these problems we shall have to apply Coulomb's laws *locally* to infinitesimal areas of contact and to integrate the results. To do this, we must ascertain the distribution of the normal force at the contact surface, an undertaking which is usually difficult and well beyond the capabilities of rigid-body statics, as explained earlier. However, we can at times compute frictional effects approximately by *estimating* the manner of distribution of the normal force at the surface of contact.

EXAMPLE 3.19

Compute the frictional resistance to rotation of a dry thrust bearing maintaining a load P as shown in Fig. 3.74.

This clearly is no longer a simple distribution as far as direction of

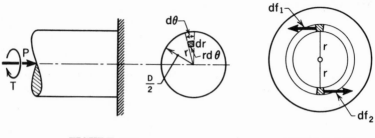

FIGURE 3.74   FIGURE 3.75

friction forces is concerned. We therefore take an infinitesimal area for examination. This area is shown in Fig. 3.74, where the element has been chosen to be related simply to the boundaries. The area dA is equal to $r\,d\theta\,dr$. We shall assume that the normal force is uniformly distributed over the area of contact, and this will permit us to get an approximate result. The normal force on the area element is then:

$$dN = \left(\frac{P}{\pi D^2/4}\right) r\,d\theta\,dr$$

The friction force associated with this force during motion is:

$$df = \mu_d \frac{P}{\pi D^2/4} r\,d\theta\,dr$$

The direction of df must oppose the relative motion between the surfaces. The relative motion is along concentric circles about the centerline, so the direction of a force df_1 (Fig. 3.75) must lie tangent to a circle of radius r. At 180° from the position of the area element for df_1, we may carry out a similar calculation for a force df_2, which for the same r must be equal and opposite to df_1, thus forming a couple. Since the entire area may be decomposed this way, we can conclude that there are only couples in the plane of contact. If we take moments about the center, we get the magnitude of the total frictional couple. The direction of the moment representation of the couple is along the shaft axis. First consider area elements on the ring of radius r:

$$dM = \int_0^{2\pi} \mu_d \frac{rP}{\pi D^2/4} r\,d\theta\,dr$$

Taking μ_d as constant, we have on integration:

$$dM = \mu_d \frac{P}{\pi D^2/4} 2\pi r^2\,dr$$

We must integrate between $r = 0$ and $r = D/2$ to get the frictional torque. Thus:

$$M = \mu_d \frac{8P}{D^2} \int_0^{D/2} r^2 \, dr = \frac{PD\mu_d}{3}$$

Other assumptions about normal force distributions at the contact surface are possible and they lead in some cases to more accurate results. For instance, if the bearing is somewhat worn, there will be a tendency for the contact surface to be slightly conical in shape. We may then assume that the pressure is greatest at the center and varies inversely as we go out. That is, the pressure distribution p can be approximated as:

$$p = \frac{a}{r}$$

where a is a constant that we can determine by summing forces on the shaft in the direction of the thrust. Thus:

$$P = \int_0^{D/2} \int_0^{2\pi} \frac{a}{r} \, r \, dr \, d\theta$$

$$P = 2\pi a \int_0^{D/2} dr = 2\pi r a \bigg]_0^{D/2} = \pi D a$$

Hence:

$$a = \frac{P}{\pi D}$$

We may now proceed as before. That is:

$$dN = \frac{P}{\pi D} \, d\theta \, dr$$

$$df = \mu_d \frac{P}{\pi D} \, d\theta \, dr$$

You may want to carry out the remainder of this problem in order to compare the result for this assumption with the case of the uniform distribution.

EXAMPLE 3.20

In Fig. 3.76 we have a cylinder B fitting over and rotating at constant angular speed about a conical body C. The cylinder supports a force P as shown in the diagram. Taking the coefficient of dynamic friction as μ_d, compute the resisting torque T for this motion. Assume that the pressure at the contact surface is uniformly distributed.

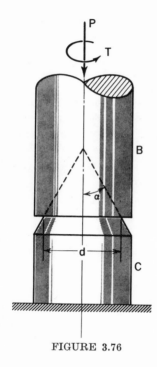

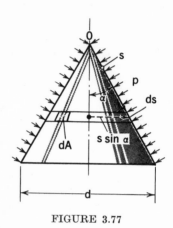

FIGURE 3.76   FIGURE 3.77

We first compute the pressure on the inside conical contact surface (see Fig. 3.77). For convenience, we shall use the distance s from O along the surface as the independent variable. A slice of the cone of length ds is shown in the figure. The vertical force from pressure p on this slice is then:

$$dF = (p)[2\pi (s \sin \alpha) \, ds](\sin \alpha) \tag{a}$$

Integrating from $s = 0$ to $s = d/(2 \sin \alpha)$, we then have the total vertical force on the contact surface. Thus:

$$F = \int_0^{d/(2 \sin \alpha)} p 2\pi \sin^2 \alpha \, s \, ds = 2\pi p \sin^2 \alpha \left. \frac{s^2}{2} \right]_0^{d/(2 \sin \alpha)}$$

$$\therefore F = p \frac{\pi d^2}{4} \tag{b}$$

From equilibrium considerations, it is clear that $F = P$ where P is the given applied load. Solving for p we then have:

$$p = \frac{4P}{\pi d^2} \tag{c}$$

The friction force df on an elemental area dA (see Fig. 3.77) is then, according to Coulomb's law,

$$df = \mu_d p \, dA = \mu_d \frac{4P}{\mu d^2} \, dA \qquad \textbf{(d)}$$

Using the peripheral area of the elemental slice shown in the figure, we can express the differential of the desired torque T as follows:

$$dT = (s \sin \alpha)\left(\mu_d \frac{4P}{\pi d^2}\right)(2\pi s \sin \alpha \, ds) \qquad \textbf{(e)}$$

Integrating over the entire contact surface, we get:

$$T = \int_0^{d/(2\sin\alpha)} \frac{8P\mu_d \sin^2 \alpha}{d^2} s^2 \, ds$$

$$T = \left(\frac{8P\mu_d \sin^2 \alpha}{d^2}\right)\left(\frac{d^3}{24 \sin^3 \alpha}\right) = \frac{P\mu_d d}{3 \sin \alpha} \qquad \textbf{(f)}$$

If α becomes 90° we get back to the flat thrust bearing examined in the previous example.

---

*EXAMPLE 3.21

A 500-lb rod is pinned at B (Fig. 3.78). What is the smallest tension T that is needed to keep the rod in the position shown? The rod is 25 ft in length

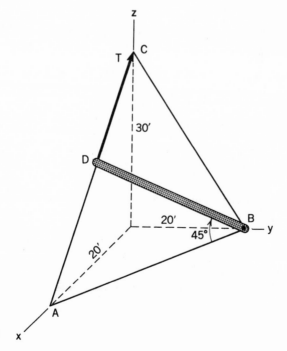

FIGURE 3.78

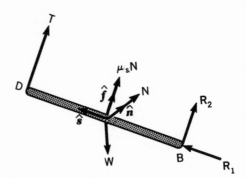

FIGURE 3.79

and has a static coefficient of friction with the inclined surface ABC of 0.3. Assume that the contact pressure is uniformly distributed.

In Fig. 3.79 we have shown a free-body diagram of the rod. The friction force is at the center of the rod in a direction normal to the rod in the plane ABC. The unit vectors $\hat{n}, \hat{f},$ and $\hat{s}$ have been shown in this diagram, since they will be of use in subsequent calculations. We may compute $\hat{n}$ by first formulating the area vector $\overrightarrow{ABC}$ as follows:

$$\overrightarrow{ABC} = \tfrac{1}{2}(30)(20)\boldsymbol{i} + \tfrac{1}{2}(20)(30)\boldsymbol{j} + \tfrac{1}{2}(20)(20)\boldsymbol{k}$$
$$\overrightarrow{ABC} = 300\boldsymbol{i} + 300\boldsymbol{j} + 200\boldsymbol{k}$$

Accordingly, we have for $\hat{n}$:

$$\hat{n} = \frac{\overrightarrow{ABC}}{ABC} = \frac{300}{ABC}\boldsymbol{i} + \frac{300}{ABC}\boldsymbol{j} + \frac{200}{ABC}\boldsymbol{k}$$
$$\hat{n} = 0.640\boldsymbol{i} + 0.640\boldsymbol{j} + 0.427\boldsymbol{k} \tag{a}$$

To compute $\hat{s}$, we shall need the position of point D. We can get the distance AD using the law of cosines for triangle ADB in Fig. 3.78. Thus:

$$AD = [(20^2 + 20^2) + 25^2 - 2(25)(\sqrt{20^2 + 20^2})\cos 45°]^{1/2}$$
$$\therefore AD = 20.6 \text{ ft}$$

The coordinates of point D are now readily computed:

$$x_D = 20 - AD\frac{20}{\sqrt{20^2 + 30^2}} = 8.57$$

$$y_D = 0$$

$$z_D = AD\frac{30}{\sqrt{20^2 + 30^2}} = 17.16$$

The vector $\hat{s}$ now may be computed:

$$\hat{s} = \frac{r_D - r_B}{|r_D - r_B|} = \frac{(8.57i + 17.16k) - 20j}{\sqrt{8.57^2 + 17.16^2 + 20^2}}$$

$$\hat{s} = 0.310i - 0.723j + 0.620k \tag{b}$$

Now using the equations of equilibrium we first determine N by summing forces in the direction $\hat{n}$. Thus:

$$N\hat{n} \cdot k = W$$

$$\therefore N = 1170 \text{ lb} \tag{c}$$

Now taking moments about an axis at B in the $\hat{n}$ direction, we get:

$$(25\hat{s} \times T) \cdot \hat{n} + [12.5\hat{s} \times (-500k)] \cdot \hat{n} + (0.3N)(12.5) = 0 \tag{d}$$

Noting that T is:

$$T = T\frac{r_C - r_A}{|r_C - r_A|} = T\frac{30k - 20i}{\sqrt{30^2 + 20^2}}$$

$$= T(-0.555i + 0.832k) \tag{e}$$

and substituting the rectangular components for the vectors $\hat{s}$, T, and $\hat{n}$ in Eq. (d), we arrive directly at the following result for T:

$$T = 359 \text{ lb} \tag{f}$$

Problems

▶ In the following problems assume pins and sockets to be friction-less. Neglect weight of members unless otherwise indicated.

1. [3.1] Draw complete free-body diagrams for the member AB and for cylinder D in Fig. 3.80. Neglect friction at the contact surfaces of the cylinder. The weights of the cylinder and the member are denoted as W_D and W_{AB} respectively.

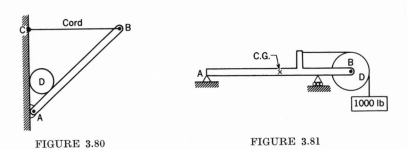

FIGURE 3.80   FIGURE 3.81

2. [3.1] Draw a free-body diagram of the beam AB and the pulley D in Fig. 3.81. The weight of the pulley is W_D and the weight of the beam is W_{AB}.

3. [3.1] Draw a free-body diagram for each member of the system shown in Fig. 3.82. Neglect weights of the members. Replace distributed load by resultant.

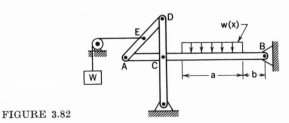

FIGURE 3.82

4. [3.1] Draw free-body diagrams of the plate $ABCD$ and the bar EG in Fig. 3.83. Assume there is no friction at the pulley H or at the contact surface C.

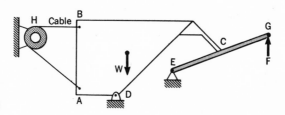

FIGURE 3.83

5. [3.2] Make a free-body diagram of the portion of the beam that is exposed from the wall in Fig. 3.84. Replace all distributions by simpler equivalent force systems. Neglect weight of the beam.

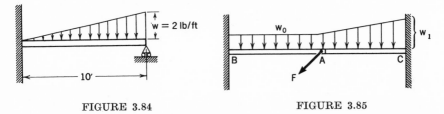

FIGURE 3.84   FIGURE 3.85

6. [3.2] Two cantilever beams are pinned together at A in Fig. 3.85. Draw free-body diagrams of each cantilever beam.

7. [3.2] Draw the free-body diagram for the structure ABC shown in Fig. 3.86. Now "cut" the structure along MM as shown by the dashed line and give the free-body diagrams for the left-hand and right-hand portions of the structure. Do not neglect the weight of members. Assume each member has its center of gravity at its midpoint.

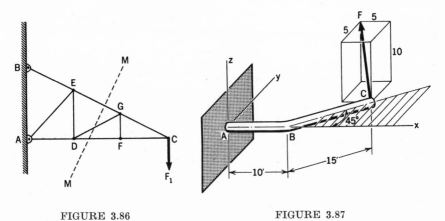

FIGURE 3.86   FIGURE 3.87

8. [3.2] Draw the free-body diagram of the bent cantilevered beam shown in Fig. 3.87. Use only xyz components of all vectors drawn.

9. [3.2] Draw a free-body diagram of the structure shown in Fig. 3.88. At E and F we have ball joint connections, while at A and G we have the members embedded in a concrete support. Use xyz components of all vectors drawn.

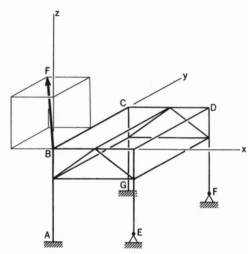

FIGURE 3.88

10. [3.2] Draw free-body diagrams of bars CD and AB in Fig. 3.89. Take the weight of the members as W_{CD} and W_{AB} respectively.

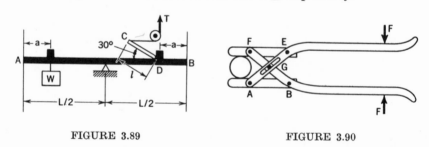

FIGURE 3.89   FIGURE 3.90

11. [3.2] Draw the free-body diagram of the component parts in Fig. 3.90. Neglect weights of all members.

12. [3.4] Consider a concurrent system of forces in space. Show that setting the sum of the forces equal to zero is equivalent to setting the sum of the moments of the forces about some point in space equal to zero.

13. [3.4] Explain why having $\sum_i (F_y)_i = 0$, $\sum_i (M_d)_i = 0$, and $\sum_i (M_e)_i = 0$ as discussed in the footnote of p. 98 is sufficient for equilibrium.

14. [3.5] Find the tensile force in cables AB and CB shown in Fig. 3.91. The remaining cables ride over frictionless pulleys E and F.

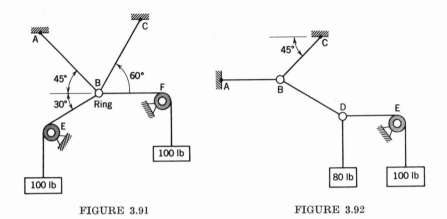

FIGURE 3.91   FIGURE 3.92

15. [3.5]) Find the force transmitted by wire BC shown in Fig. 3.92. The pulley E can be assumed to be frictionless in this problem.

16. [3.5] Cylinders A and B weigh 500 lb each in Fig. 3.93 and cylinder C weighs 1000 lb. Compute all contact forces.

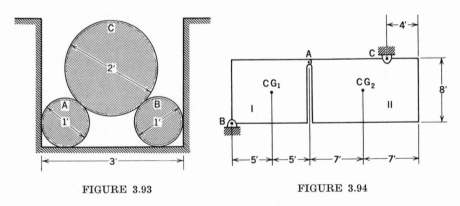

FIGURE 3.93   FIGURE 3.94

17. [3.5] Find the components of the forces acting on pins A, B, and C connecting and supporting the blocks shown in Fig. 3.94. Block I weighs 10 tons and block II weighs 30 tons.

18. [3.5] What load W will a pull P of 100 lb lift in the pulley system shown in Fig. 3.95? Sheaves A, B, and C weigh 20 lb, 15 lb, and 30 lb, respectively.

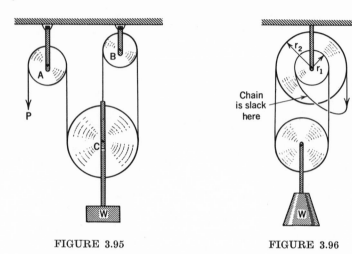

FIGURE 3.95   FIGURE 3.96

19. [3.5] Shown in Fig. 3.96 is a *differential pulley*. Compute F in terms of W, r_1, and r_2.

20. [3.5] Shown in Fig. 3.97 is a scenic excursion train with cog wheels for steep inclines. The train loaded weighs 30 tons. If the cog wheels have a mean radius to the contact points of the teeth of 2 ft, what torque must be applied to the driver wheels A if wheels B run free? What force do wheels B transmit to the ground?

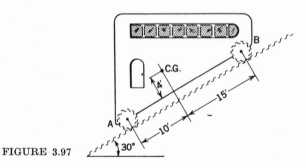

FIGURE 3.97

21. [3.5] What is the largest weight W that the crane in Fig. 3.98 can lift without tipping? What are the supporting forces when the crane lifts this load? What is the force system transmitted through section C of the beam? Compute the force transmitted through section D. The crane weighs 10 tons having a center of gravity as shown in the diagram.

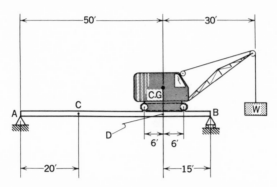

FIGURE 3.98

22. [3.5] Shown in Fig. 3.99 is a bent-lever balance in the process of measuring a weight of 25 lb, which causes the balance to swing 30° as shown. If the arm and ball AB weigh 10 lb and we neglect the weight of arm CA, what angle will be formed by a 30-lb weight? Neglect the weight of the container D.

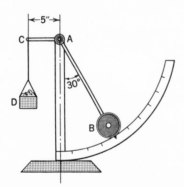

FIGURE 3.99

23. [3.5] A beam of length 20 ft weighing 800 lb supports an 800-lb load at the left end, a 1000-lb force 12 ft from the left end, and a 2-ton load at the right end. The left support is positioned 5 ft from the left end and is capable of supporting a 2000-lb load. How far from the right end must the second support be positioned to load the left end to one-half its capacity?

24. [3.5] A speedboat drags a water skier and kite as shown in Fig. 3.100. The towline has a tension at the kite of 200 lb. What is the vertical lift of the kite for steady elevation of the skier? Weight of skier is 180 lb.

FIGURE 3.100

25. [3.5] A 30-ton tank (Fig. 3.101) is climbing up a 30° incline at constant speed. What is the torque developed on the rear drive wheels to accomplish this? Assume all other wheels are free-turning.

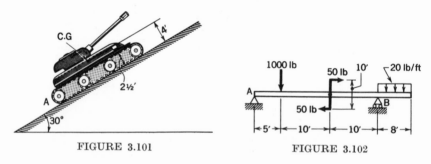

FIGURE 3.101

FIGURE 3.102

26. [3.5] In Fig. 3.102, find the supporting forces at A and B on the beam, whose weight is 500 lb. Is the location of the couple significant for this calculation? Why?

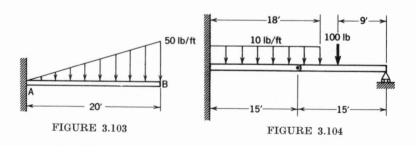

FIGURE 3.103

FIGURE 3.104

27. [3.5] Ascertain the resultant force system being transmitted through the cross section of the beam at A in Fig. 3.103. The beam weighs 200 lb. Could you make such a calculation if this beam were also supported at B? Why?

28. [3.5] Compute the supporting forces for the cantilever beam and pinned extension shown in Fig. 3.104.

29. [3.5] Find the supporting force system for the cantilever beams shown in Fig. 3.105 connected to bar AB by pins.

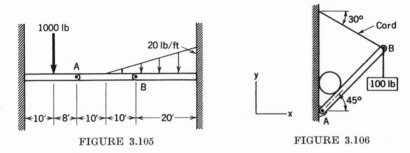

FIGURE 3.105   FIGURE 3.106

30. [3.5] A cylinder having a diameter of 4 ft and a weight of 200 lb is supported by beam AB of length 20 ft and weight 50 lb as shown in Fig. 3.106. If the surfaces of contact of the cylinder are frictionless, determine the supporting force components in the x and y directions at A.

31. [3.5] In Fig. 3.107, the pulley at D weighs 500 lb. Neglecting the weights of the bars, find the force transmitted from one bar to the other at C.

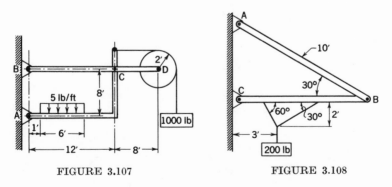

FIGURE 3.107   FIGURE 3.108

32. [3.5] Solve for the supporting forces at A and C in Fig. 3.108. AB weighs 100 lb and BC weighs 150 lb.

33. [3.5] What are the forces in the cables shown in Fig. 3.109 supporting a 100-lb weight? Note that cable BD lies in the yz plane.

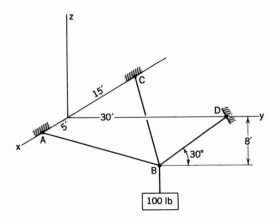

FIGURE 3.109

34. [3.5] A vertical cable AB and two other inclined cables BD and CB restrain the 100-lb force shown in Fig. 3.110. Compute the tensile forces in the cables.

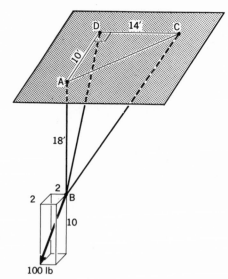

FIGURE 3.110

35. [3.5] What is the force system transmitted through section A for the cantilever beam of Prob. 8? Compute the force system transmitted through section B. Take $F = 1000$ lb.

36. [3.5] Compute the value of F to maintain the 200-lb weight shown in Fig. 3.111. Assume the bearings are frictionless and thin and determine the forces from the bearings on the shaft at A and B.

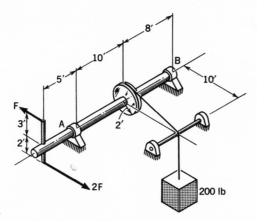

FIGURE 3.111

37. [3.5] Find the supporting force system for the cantilever beam shown in Fig. 3.112. What is the force system transmitted through a cross section of the beam at *B*?

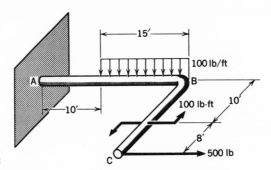

FIGURE 3.112

38. [3.5] Shown in Fig. 3.113 is a bar with two right-angle bends and supporting a force *F* given as:

$$F = 10\boldsymbol{i} + 3\boldsymbol{j} + 100\boldsymbol{k}\ \text{lb}$$

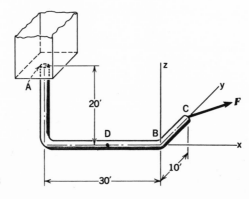

FIGURE 3.113

If the bar has a weight of 10 lb per foot, what is the supporting force system at A?

39. [3.5] Do Prob. 38 for the case where there is additionally a parabolic loading in the vertical direction along BC given as:

$$w_z = 6y^2 \text{ lb/ft}$$

What is the force system transmitted through the cross section of the beam at position D, which is 15 ft to the left of B?

40. [3.5] What are the supporting forces for the beam shown in Fig. 3.114 having two right-angle bends? The beam weighs 20 lb/ft and is acted on by a force F given as:

$$F = -100i + 500j - 600k \text{ lb}$$

What is the force system transmitted at E? A, B, and F are ball joints.

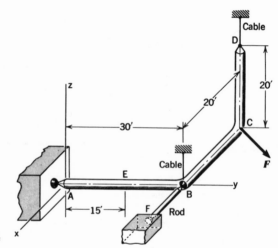

FIGURE 3.114

41. [3.5] Shown in Fig. 3.115 is a structure supported by a socket joint at A, a pin connection at B offering no resistance in the direction AB, and a simple roller support at C. What are the supporting forces for the loads shown?

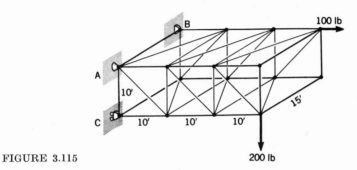

FIGURE 3.115

42. [3.5] Shown in Fig. 3.116 is a transport plane having a gross weight of 70,000 lb with a center of gravity as shown in the diagram. Wheels A and B are locked by the braking system while an engine is being tested under load prior to take off. A thrust T of 3000 lb is developed by this engine. What are the supporting forces?

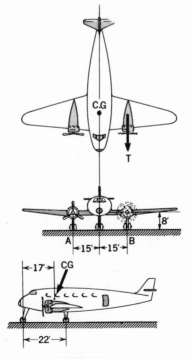

FIGURE 3.116

43. [3.5] Shown in Fig. 3.117 is a blimp fixed at the mooring tower at D and held by cables AB and AC. The blimp weighs 3000 lb. The resultant force from air pressure and winds is shown as components of vector F. If $F = 3500i + 200j + 300k$ lb, compute the tension in the cables. What force system is transmitted to the ground at F through the mooring tower?

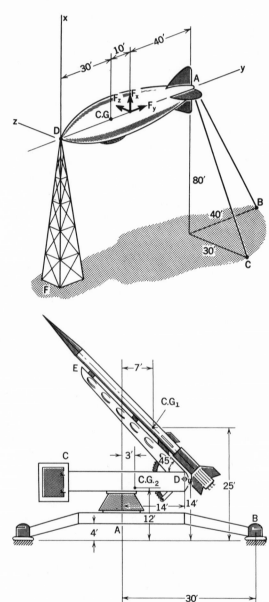

FIGURE 3.117

FIGURE 3.118

44. [3.5] Shown in Fig. 3.118 is a 10-ton sounding rocket (used for exploring outer space) having a center of gravity shown as C.G.$_1$. It is mounted on a launcher whose weight is 50 tons with a center of gravity at C.G.$_2$. The launcher has three identical legs separated 120° from each other. Leg AB is in the same plane as the rocket and supporting arms CDE. What are the supporting forces from the ground? What torque is transmitted from the horizontal arm CD to the ramp ED at hinge D to counteract the load of the rocket?

45. [3.5] Assume the rocket in Prob. 44 has been ignited and is developing a thrust of 15,000 lb. Compute the normal components of the supporting forces and the torque transmitted at point D when the rocket has moved 1 ft and is moving at this position with constant speed.

46. [3.5] What force P is needed to hold the door in Fig. 3.119 in a horizontal position? The door weighs 50 lb. Determine the supporting forces at A and B. At A there is a pin (neglect moments) and at B there is a ball joint.

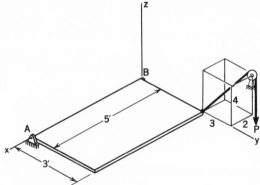

FIGURE 3.119

47. [3.6] Find the horizontal force developed by the rock crusher shown in Fig. 3.120 at A. The pressure p is 70 psi above atmosphere. Rods AB, BD,

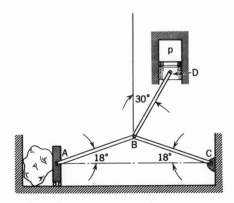

FIGURE 3.120

and BC can be considered weightless for this problem. Diameter of the piston is 1 ft.

48. [3.6] Do Prob. 47 in the case where an additional 3000-lb vertical force is applied at the midpoint of the rod AB.

49. [3.6] Find the force delivered at C in a horizontal direction to crush the rock (Fig. 3.121). Pressure $p_1 = 100$ psi and $p_2 = 60$ psi (pressures measured above atmospheric pressure). The diameters of the pistons are 6 in. each. Neglect the weight of the rods.

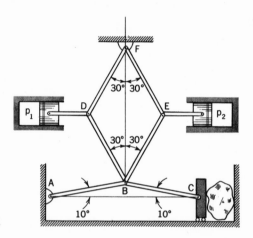

FIGURE 3.121

50. [3.6] Find the forces in the cables DB and CB as well as the compression member AB shown in Fig. 3.122. The 500-lb force is parallel to the y axis.

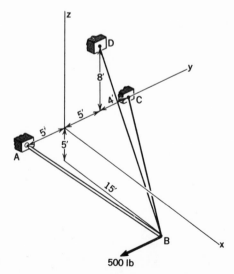

FIGURE 3.122   500 lb

51. [3.6] In Fig. 3.123 is shown a block of material weighing 200 lb. It is supported by members KC and HB, whose weight we neglect, a ball-socket joint support at A, and a smooth frictionless support at E. Members KC and HB have directions collinear with diagonals of the block as shown. What are the supporting forces for this block?

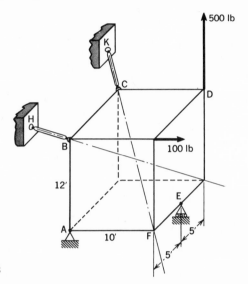

FIGURE·3.123

52. [3.6] In Fig. 3.124 is shown a trap door kept open by a rod CD, whose weight we shall neglect. The door has hinges at A and B and has a weight of 200 lb. A wind blowing against the outside surface of the door creates a pressure increase of 2 lb/ft². Find the force in the rod, assuming it cannot slip from the position shown. Also determine the forces transmitted to the hinges. Only hinge B can resist motion along direction AB.

▶ Problems 53–71 provide a mixture of exercises encompassing the range of material covered in Sections 3.1 through 3.6.

FIGURE 3.124   FIGURE 3.125

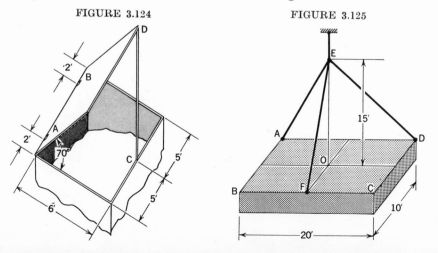

53. [3.6] In Fig. 3.125, if point O is at the center of the upper face of the block weighing 1000 lb, find the tension in the cords AE, ED, and EF.

54. [3.6] Find the supporting forces at the socket connections A, D, and C in Fig. 3.126. Members AB and DB are pinned together through member EC at B.

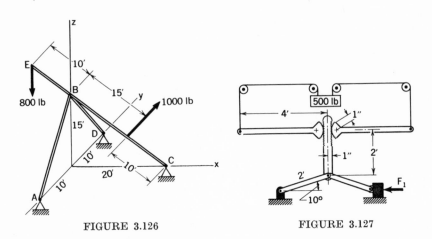

FIGURE 3.126   FIGURE 3.127

55. [3.6] What force F_1 will be developed by the 500-lb load as shown in Fig. 3.127? Neglect friction. The design is symmetrical.

56. [3.6] Find the forces on the block of ice from the hooks at A and F in Fig. 3.128.

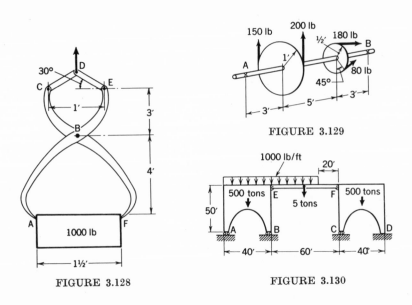

FIGURE 3.128   FIGURE 3.130

57. [3.6] Find the forces transmitted to the bearings shown at A and B in Fig. 3.129. The forces on the discs are from belts. The weight of the larger pulley is 30 lb and that of the smaller one 20 lb. The shaft weight is 50 lb.

58. [3.6] Find the supporting forces on the beam EF and the supporting forces at A, B, C, and D (Fig. 3.130).

59. [3.6] A bar can rotate parallel to plane A about an axis of rotation normal to the plane at O (Fig. 3.131). A weight W is held by a cord which is attached to the bar over a pulley that can rotate as the bar rotates. Find the relation between the couple C and ϕ for equilibrium.

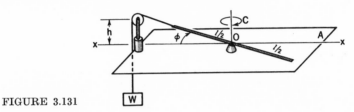

FIGURE 3.131

60. [3.6] Neglecting friction, find the angle β of line AB for equilibrium in Fig. 3.132.

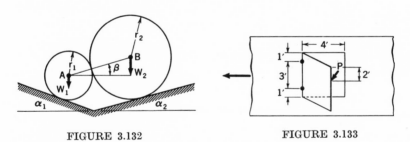

FIGURE 3.132   FIGURE 3.133

61. [3.6] In Fig. 3.133, determine the force P required to keep the 30-lb door of an airplane open 30° while in flight. The force P is exerted in a direction normal to the fuselage. There is a net pressure increase on the outside surface of 3 lb/in.². Also determine the supporting forces at the hinges.

62. [3.6] What is the resultant of the force system transmitted across the section at *A* in Fig. 3.134? Couple is parallel to plane *M*.

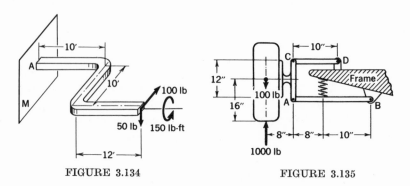

FIGURE 3.134   FIGURE 3.135

63. [3.6] The pavement exerts a force of 1000 lb on the tire as shown in Fig. 3.135. The tire, brakes, etc. weigh 100 lb; the center of gravity is taken at the center plane of the tire. Determine the force from the spring and the compression force in *CD*.

64. [3.6] Find the supporting forces at *A* and *B* (Fig. 3.136).

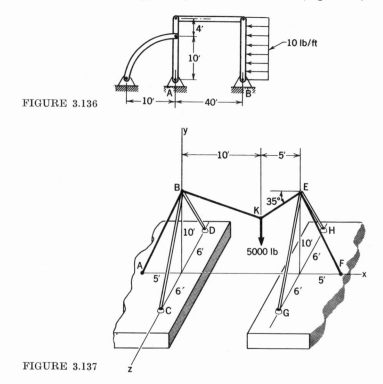

FIGURE 3.136

FIGURE 3.137

65. [3.6] Determine the supporting forces at A, C, D, G, F, and H for the system shown in Fig. 3.137.

66. [3.6] In Fig. 3.138, what change in elevation for the 100-lb weight will a couple of 300 lb-ft support if we neglect friction in the bearings at A and B? Also determine the supporting force components at the bearings.

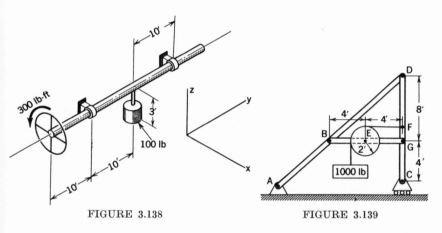

FIGURE 3.138   FIGURE 3.139

67. [3.6] Determine the force components at G in Fig. 3.139.

68. [3.6] A beam weighing 400 lb is held by a socket joint at A and by two cables CD and EF (Fig. 3.140). Find the tension in the cables. They are attached at opposite ends of the beam as shown.

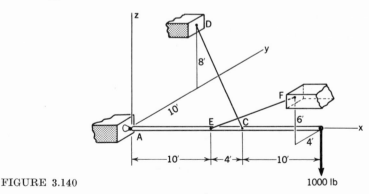

FIGURE 3.140   1000 lb

69. [3.6] Shown in Fig. 3.141 is a small helicopter in a hovering maneuver. The helicopter rotor blades give a lifting force F_1 but there results from the air forces on the blades a torque C_1. The rear rotor prevents the helicopter from rotating about the z axis. Compute the force F_1 and couple C_2 in terms of the weight W. How are F_3 and C_1 related?

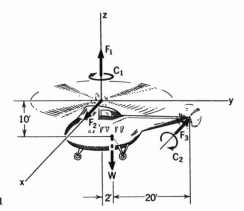

FIGURE 3.141

70. [3.6] Assume that the wheel shown in Prob. 63 comes to rest on a rock and that the "knee action" causes the frame to remain horizontal and AB to rotate 20°. Compute the compressive force on member CD.

*71. [3.6] Shown in Fig. 3.142 is a *walking beam* engine used on steamers in the 1840's. Power comes from a main cylinder G and is transmitted through connecting rods ef and bc, walking beam cde, and crank ab which is connected to the paddle wheel W through a shaft. What is the moment on W at the instant when crank ab has the position shown in the figure? Assume the steam pressure p is 100 psi at this position.

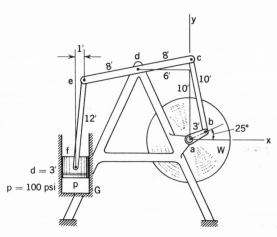

FIGURE 3.142

72. [3.8] In Fig. 3.143, find the forces transmitted by each member. Be sure to state whether they are tension or compression members.

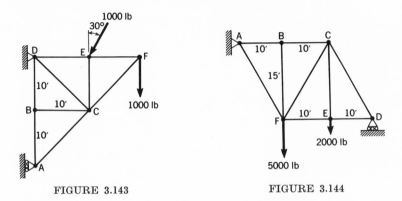

FIGURE 3.143   FIGURE 3.144

73. [3.8] In Fig. 3.144, find the forces transmitted by each member.

74. [3.8] Find the forces transmitted by the straight members in Fig. 3.145.

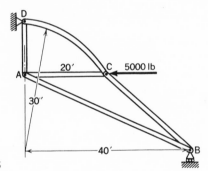

FIGURE 3.145

75. [3.8] In Fig. 3.146 find the force in each of the members in terms of kips (1 kip = 1000 lb).

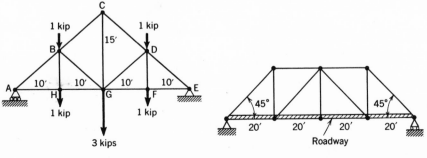

FIGURE 3.146   FIGURE 3.147

76. [3.8] The bridge truss shown in Fig. 3.147 supports a roadway load of 1000 lb/ft. Each member weighs 30 lb/ft. Compute the forces in the members, accounting approximately for the weight of the members.

77. [3.8] Find the force in members HE, FH, FE, and FC of the truss shown in Fig. 3.148.

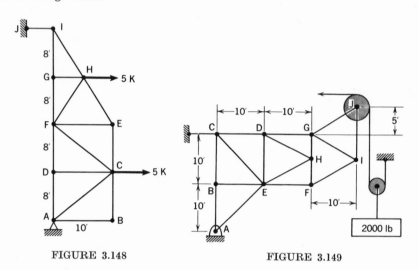

FIGURE 3.148   FIGURE 3.149

78. [3.8] Find the force in members FI, EF, and DH in the truss shown in Fig. 3.149. Neglect the weight of the pulleys.

79. [3.8] (a) In Fig. 3.150, find the forces transmitted by member DC. (b) What is the force transmitted by DE?

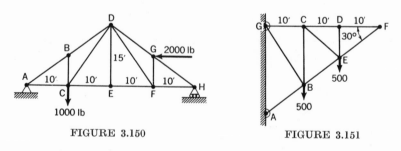

FIGURE 3.150   FIGURE 3.151

80. [3.8] Find the forces in members CB and BE in Fig. 3.151.

81. [3.8] Determine the force transmitted by member KN in the plane truss shown in Fig. 3.152.

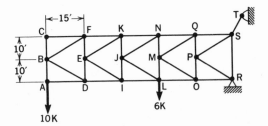

FIGURE 3.152

82. [3.8] Shown in Fig. 3.153 is a truss supporting a roadway load of 800 lb/ft. Concentrated loads have been shown representing approximations of vehicle loading at some instant of time. The bridge has six 20-ft panels. Determine the forces in members EG, FH, and IJ.

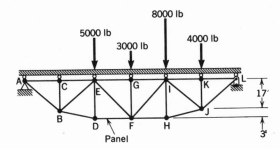

FIGURE 3.153   Panel

83. [3.8] Find the forces in member JF in Fig. 3.154.

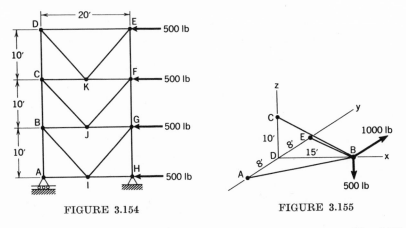

FIGURE 3.154   FIGURE 3.155

84. [3.8] Find the forces in the members in Fig. 3.155. The 1000 lb force is parallel to y.

85. [3.8] In Fig. 3.156, find the forces in the members and the supporting forces for the space truss *ABCD*. Note that *BDC* is in the *xz* plane.

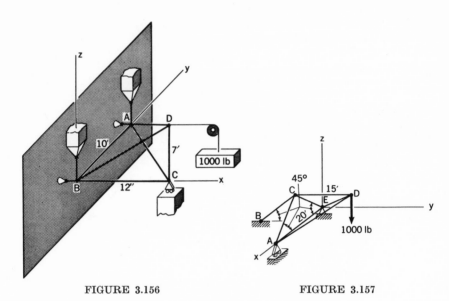

FIGURE 3.156   FIGURE 3.157

86. [3.8] In Fig. 3.157, find the forces in all the members. Note that *ACE* is in the *xz* plane.

87. [3.8] Find the forces in the members of the space truss shown in Fig. 3.158 under the action of a force *F* given as:

$$F = 10i - 6j - 12k \text{ kips}$$

Note that *C* is a ball-socket joint while *A*, *F*, and *E* are on rollers.

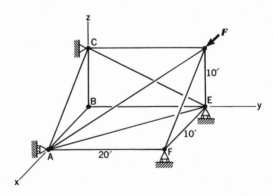

FIGURE 3.158

88. [3.8] In Fig. 3.159 the plane of ball sockets $CDHE$ is in the zy plane while the plane of $FGDE$ is parallel to the xz plane. Note that this is *not* a simple space truss. Nevertheless, the forces in the members can be ascertained by choosing a desirable starting joint and proceeding by statics from joint to joint. Determine the forces in all the members and then determine the supporting forces.

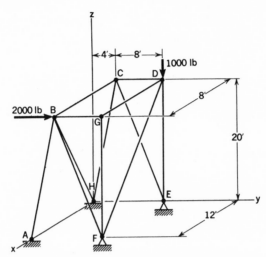

FIGURE 3.159

89. [3.9] If the coefficient of friction is $\mu_s = 0.3$, find the largest angle θ before the block begins to slide down the incline in Fig. 3.160.

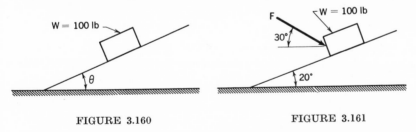

FIGURE 3.160   FIGURE 3.161

90. [3.9] In Fig. 3.161, what is the value of the force F, inclined at 30° to the horizontal, needed to get the block just started up the incline? What is the force F needed to keep it just moving up at a constant speed? The coefficients of static and dynamic friction are 0.3 and 0.275, respectively.

91. [3.9] A block has a force F applied as shown in Fig. 3.162. If this force has a time variation as shown in the diagram, draw a simple sketch showing the friction force variation with time. Take $\mu_s = 0.3$ and $\mu_d = 0.2$ for the problem.

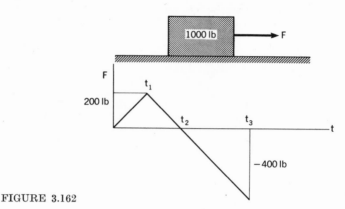

FIGURE 3.162

92. [3.9] Find the force F needed to start the 200-lb weight (Fig. 3.163) moving to the right, if the coefficient of friction is $\mu_s = 0.35$.

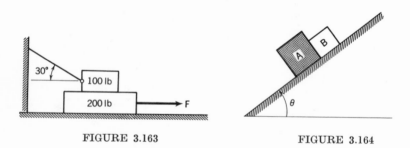

FIGURE 3.163   FIGURE 3.164

93. [3.9] Show that by increasing the inclination ϕ on an inclined surface until there is impending slippage of supported bodies, we reach the *angle of repose* ϕ_s, so that

$$\tan \phi_s = \mu_s$$

94. [3.9] Bodies A and B in Fig. 3.164 weigh 500 lb and 300 lb, respectively. The platform on which they are placed is raised from the horizontal position to an angle θ. What is the maximum angle that can be reached before the bodies slip down the incline? Take μ_s for body B and the plane as 0.2 and μ_s for body A and the plane as 0.3.

95. [3.9] A car travels at 60 mph along a plane circular path of radius R. If the coefficient of friction between tires and road is 0.4, what is the smallest turning radius of the car before skidding occurs. We assume that the weight of the car is carried equally by the two tires on the outer side of the curve? Take the weight of the car as 3000 lb. The centrifugal force on the car, you will recall from physics, is mV^2/R.

96. [3.9] A homogeneous rod weighing 100 lb rests at a corner as shown in Fig. 3.165. What is the minimum angle possible before slipping occurs? The coefficient of static friction at A is 0.3 and at B is 0.2.

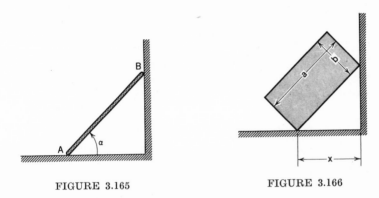

FIGURE 3.165   FIGURE 3.166

97. [3.9] A homogeneous block of weight W is to be placed at a corner as is shown in Fig. 3.166. Set up an equation for determining the maximum distance x away from the wall before slipping occurs. The coefficient of static friction is the same for all surfaces of contact.

98. [3.9] In Fig. 3.167 is a tank weighing 500 lb containing water as shown. The tank is to be moved by a force F on each of the parallel connecting rods. If $\mu_s = 0.2$, what force is required to move this tank along the floor?

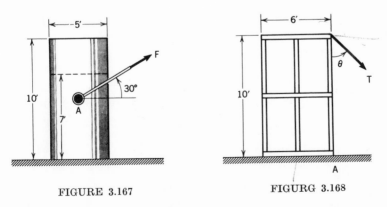

FIGURE 3.167   FIGURG 3.168

99. [3.9] A loaded crate is shown in Fig. 3.168. The crate weighs 500 lb with a center of gravity at its geometric center. The contact surface between crate and floor has a static coefficient of friction of 0.2. If $\theta = 90°$, show that the crate will slide before one can increase T large enough for tipping to occur. If a stop is to be inserted in the floor at A to prevent sliding so that the crate could be tipped, what horizontal force will be exerted on the stop?

100. [3.9] In Prob. 99 compute a value of θ and T where sliding and tipping will occur simultaneously. If the actual angle θ is smaller than this value of θ, is there any further need of the stop at A to prevent sliding?

101. [3.9] A 500-lb crate A rests on a 1000-lb crate B (Fig. 3.169). The centers of gravity of the crates are at the geometric centers. The coefficients of static friction between contact surfaces are shown in the diagram. The force T is increased from zero. What is the first action to occur?

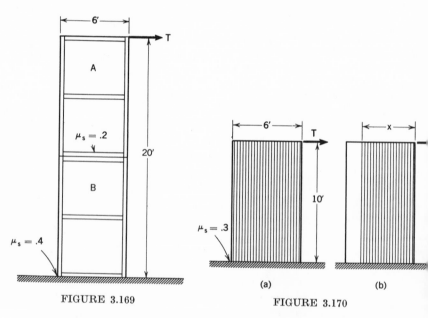

FIGURE 3.169   FIGURE 3.170

102. [3.9] A rectangular case is loaded with uniform vertical thin rods such that when it is full, as shown in Fig. 3.170 (a), the case has a total weight of 1000 lb. The case weighs 100 lb when empty and has a coefficient of static friction of 0.3 with the floor as shown in the diagram. A force T of 200 lb is maintained on the case. If the rods are unloaded as shown in Fig. 3.170 (b), what is the limiting value of x for equilibrium to be maintained?

103. [3.9] A 30-ton tank is moving up a 30° incline as shown in Fig. 3.171. If $\mu_s = 0.6$ for the contact surface between tread and ground, what torque must be developed at the rear drive sprocket? Take the mean diameter of the rear sprocket as 2 ft.

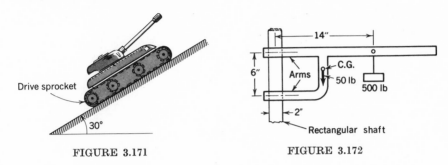

FIGURE 3.171   FIGURE 3.172

104. [3.9] What is the minimum coefficient of friction required just to maintain the bracket and its 500-lb load (Fig. 3.172) in a static position? (Assume point contacts at the centerlines of the arms.) Center of gravity is 7 in. from shaft centerline.

105. [3.9] If the coefficient of friction in Prob. 104 is 0.2, at what minimum distance from the centerline of the vertical shaft can we support the 500 lb without slipping?

106. [3.9] Given that $\mu_s = 0.2$ for all surfaces, find the force P needed to start the block A to the right in Fig. 3.173.

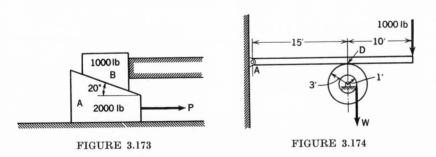

FIGURE 3.173   FIGURE 3.174

107. [3.9] In Fig. 3.174, what is the maximum load W that the 1000-lb force will hold up if the coefficient of friction μ_s at D is 0.3? Neglect all other friction and the weights of the member.

108. [3.9] In the preceding problem a weight of 300 lb is suspended from the pulley. For a load of 1000 lb, what are the normal and frictional forces transmitted to the pulley at D?

109. [3.9] In Fig. 3.175, the static and dynamic coefficients of friction for the surfaces in contact are $\mu_s = 0.3$ and $\mu_d = 0.25$. What is the minimum force P needed to just get the cylinder rolling?

110. [3.9] An insect tries to climb out of a hemispherical bowl of radius 2 ft (Fig. 3.176). If the coefficient of friction between insect and bowl is 0.4, how high up does the insect go? If the bowl is spun about a vertical axis, the bug gets pushed out in a radial direction by the force, $mr\omega^2$, as you learned in physics. At what speed ω will the bug just be able to get out of the bowl?

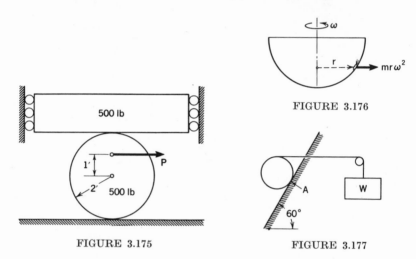

FIGURE 3.176

FIGURE 3.175

FIGURE 3.177

111. [3.9] The cylinder in Fig. 3.177 weighs 200 lb and is in equilibrium. What is the friction force at A? If there is impending slippage, what is the friction coefficient? The supporting plane is inclined at 60° to the horizontal.

112. [3.9] A rod is supported by two wheels spinning in opposite directions (Fig. 3.178). If the wheels were horizontal, the rod would be placed centrally over the wheels for equilibrium. However, the wheels have an inclination of 20° as shown, and the rod must be placed at a position off center for equilibrium. If the coefficient of friction is $\mu_D = 0.8$, how many feet off center must the rod be placed?

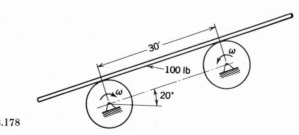

FIGURE 3.178

113. [3.9] In Fig. 3.179, the block of weight W is to be moved up an inclined plane. A rod of length c with negligible weight is attached to the block and the force F is applied to the top of this rod. If the coefficient of starting friction is μ_s, determine the maximum length c for which the block will begin to slide rather than tip.

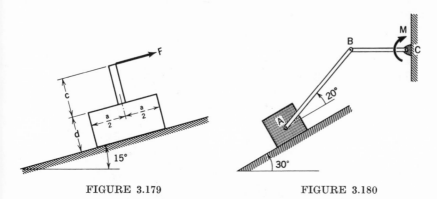

FIGURE 3.179   FIGURE 3.180

114. [3.9] What torque M is required (Fig. 3.180) to move block A weighing 100 lb up the incline? Rod AB is 30 in. in length while rod BC is 20 in. in length. Neglect the weight of these rods as well as friction in the pins. Take μ_s between block and incline as 0.3.

115. [3.9] Do Prob. 114 for the case where rod AB weighs 20 lb and rod BC weighs 15 lb. Both rods are uniform.

116. [3.9] Blocks A and B each weigh 500 lb (Fig. 3.181). What is the force P for impending motion if μ_s between blocks A and B is 0.3 and μ_s between block B and the incline is 0.4? Also determine the force in cable CD for this condition.

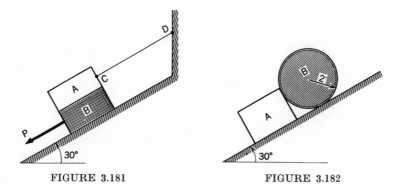

FIGURE 3.181   FIGURE 3.182

117. [3.9] A block A and cylinder B are shown as an inclined plane (Fig. 3.182). If μ_s for all contact surfaces is 0.7, will the bodies move down the plane or remain stationary? Block A weighs 200 lb and cylinder B weighs 400 lb.

118. [3.9] In Prob. 117, what is the minimum force P parallel to the incline and applied to block A such as to move the system up the incline? Take $\mu_s = 0.2$.

119. [3.9] What is the force F to hold stationary three cylinders, each weighing 100 lb (Fig. 3.183)? Take $\mu_s = 0.2$ for all surfaces of contact.

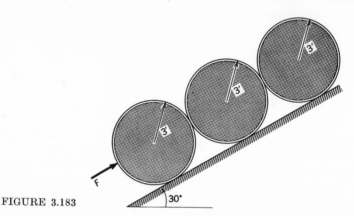

FIGURE 3.183

120. [3.9] Block C (Fig. 3.184) weighs 200 lb and identical blocks A and B weigh 150 lb each. If $\mu_s = 0.2$ for all surfaces, can the arrangement shown in the diagram remain in equilibrium?

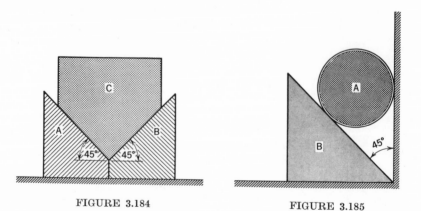

FIGURE 3.184   FIGURE 3.185

121. [3.9] A cylinder A weighing 100 lb rests on a block B weighing 500 lb as shown in Fig. 3.185. The cylinder has a diameter of 1 ft. If $\mu_s = 0.5$ for all contact surfaces, can the system stay in equilibrium as shown?

122. [3.9] What is the height x of a step shown in Fig. 3.186 so that the force P will roll the 50-lb cylinder over the step at the same time that there is impending slippage at a? Take $\mu_s = 0.3$.

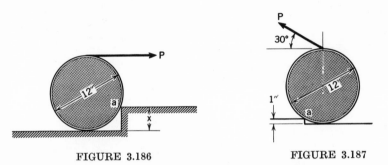

FIGURE 3.186   FIGURE 3.187

123. [3.9] Can a force P roll the 50-lb cylinder over the step (Fig. 3.187)? The coefficient of static friction is 0.4. What is the value of P if this can be done?

124. [3.9] Explain how a violin bow, when drawn over a string, maintains the vibration of the string. Do this in terms of friction forces and the difference in static and dynamic coefficients of friction.

125. [3.9] What force P is required to move the 100-lb block (Fig. 3.188)? All contact surfaces have a coefficient of static friction of 0.2. What is the tension in cord AB when this force P is applied?

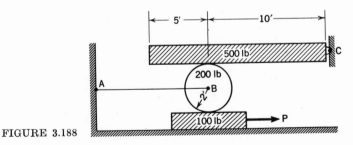

FIGURE 3.188

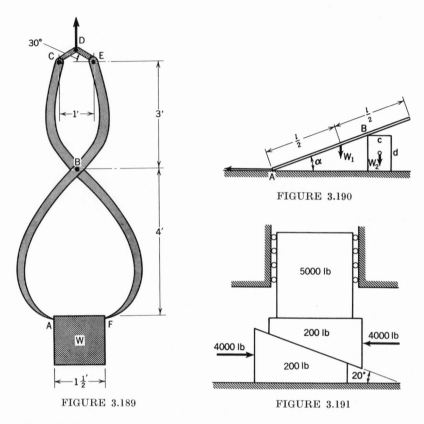

FIGURE 3.190

FIGURE 3.189   FIGURE 3.191

126. [3.9] Suppose the ice lifter of Prob. 56 is used to support a hard block of material by friction only (Fig. 3.189). What is the minimum coefficient of static friction μ_s to accomplish this for any weight W and the geometry shown in the diagram?

127. [3.9] The rod in Fig. 3.190 is pulled at A and it moves to the left. If the coefficient of dynamic friction for the rod at A and B is μ_1, what must the minimum value of W_2 be to prevent the block from tipping? With this value of W_2 determine the coefficient of static friction between the block and the supporting plane needed to just prevent the block from sliding.

128. [3.9] If we neglect friction at the rollers and if the coefficient of static friction is 0.2 for all surfaces, ascertain whether the 5000-lb weight in Fig. 3.191 will go up, down, or stay stationary.

129. [3.9] A bar rests on surfaces inclined at angle β with the horizontal (Fig. 3.192). At what maximum angle α can we place the bar without its slipping if the coefficient of friction is μ_s? Determine the supporting forces at A and B for this condition.

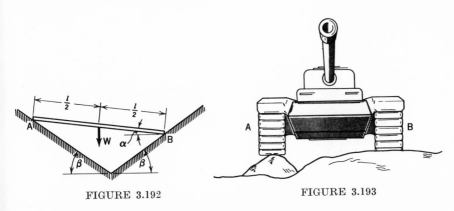

FIGURE 3.192   FIGURE 3.193

130. [3.9] A 30-ton tank shown in Fig. 3.193 is on a terrain such that one track A is supported at a small region at the center of the track while the other track B is essentially on a flat surface. The tracks are 18 inches wide and track B is in contact with the ground a distance of 20 ft. In order to turn, track B is locked and track A is powered. If the weight of the tank is equally supported by both tracks and if on track B the contact pressure is uniformly distributed, what torque must be applied to the sprocket at A to start the turning? Take $\mu_s = 0.2$ for the contact surfaces. Set up a quadrature for friction acting on the whole of the contact surface of track B. Simplify by assuming friction is concentrated along the centerline of track B. Take the distance between centerlines of track A and B as 10 ft and take the diameter of the sprocket as 2 ft.

*131. [3.9] In Fig. 3.194 is a chute having sides that are at right angles to each other. The chute is 30 ft in length with end A 10 ft higher than end B. Cylinders weighing 200 lb are to slide down the chute. What is the maximum allowable coefficient of friction so there cannot be sticking of the cylinders along the chute?

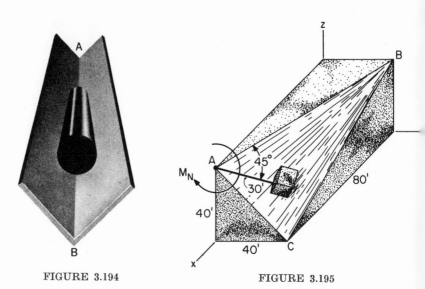

FIGURE 3.194   FIGURE 3.195

*132. [3.9] A block whose weight W is 200 lb is supported by an inclined surface ABC (see Fig. 3.195) and is guided by a rod fixed to the block and pinned at position A. The rod is 20 ft long and is to be considered weightless. Will the weight remain stationary at the position shown? If not, what minimum torque M_n is required to keep it stationary? If it is stationary, what minimum torque M_n will start it sliding down the incline? Take $\mu_s = 0.4$ and $\mu_d = 0.3$.

*133. [3.9] A body weighing 100 lb is fixed to a light rod which is pinned at C (Fig. 3.196). A force T parallel to AB and in plane ABC is applied to the body. What is the minimum value of T just to get the weight moving? Take $\mu_s = 0.4$ for the surfaces of contact.

*134. [3.9] A block rests on an inclined surface (Fig. 3.197). If the block weighs 20 lb and if the coefficient of static friction is 0.5, what is the magnitude that force P in the plane must have for impending motion? What is the direction of the impending motion?

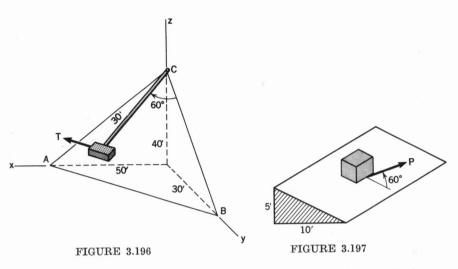

FIGURE 3.196   FIGURE 3.197

*135. [3.9] A cylinder weighing 50 lb rests on an inclined surface as shown in Fig. 3.198. If $\mu_s = 0.6$, what is the maximum value of the force P lying in plane ABC and having a direction parallel to BC that the cylinder can hold by friction? Give the unit vector for the direction of impending motion.

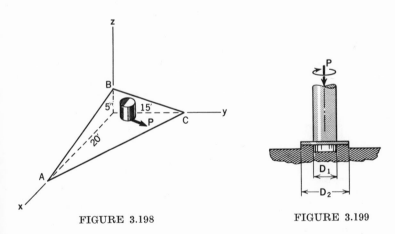

FIGURE 3.198   FIGURE 3.199

136. [3.9] Compute the frictional resisting torque for the concentric dry thrust-bearing shown in Fig. 3.199. The coefficient of friction is taken as μ_D.

137. [3.9] In Fig. 3.200 is shown the support end of a dry thrust-bearing. Four pads form the contact surface. If a shaft creates a 100-lb thrust uniformly distributed over the pads, what is the resisting torque for a coefficient of friction of 0.1?

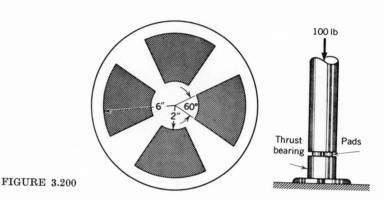

FIGURE 3.200

138. [3.9] Compute the frictional torque needed to rotate the truncated cone relative to the fixed member shown in Fig. 3.201. The cone has a 2-inch diameter base and a 60° cone angle and is cut off 0.3 in. from the cone tip. The coefficient of dynamic friction is 0.2.

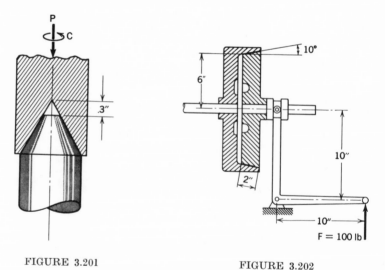

FIGURE 3.201   FIGURE 3.202

139. [3.9] Shown in Fig. 3.202 is a cone clutch. Assuming that uniform pressures exist between the contact surfaces, compute the maximum torque that can be transmitted. The coefficient of friction is 0.30 and the activating force F is 100 lb.

140. [3.9] Do Prob. 139 when the pressure increases linearly from zero at the inner part of the cone to a maximum at the outer part of the cone.

141. [3.9] A 1000-lb block is being lowered down an inclined surface (Fig. 3.203). The block is pinned to the incline at C, and at B a cord is played out so as to cause the body to rotate at uniform speed about C. Taking μ_d to be 0.3 and assuming the contact pressure is uniform along the base of the block, compute T for the configuration shown in the diagram.

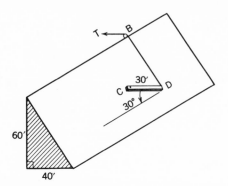

FIGURE 3.203

142. [3.9] Shown in Fig. 3.204 is an internal shoe brake. A and A' are shoes pivoted at B and B', which are considered fixed in the inertial space. Drum D with radius r and face width b revolves in a clockwise direction. What is the braking torque about the center of drum O when forces P and P' of equal magnitude are exerted by a hydraulic pump? Assume the pressure distributions between the shoes and drum are given as $p = p_0 \cos \alpha$.

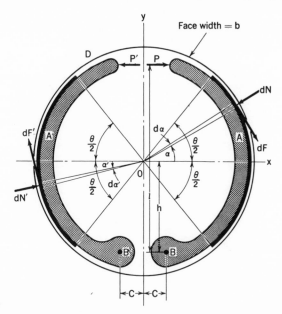

FIGURE 3.204

Appendix

VECTOR ALGEBRA REVIEW

DEFINITION 1. The *magnitude* of a vector quantity is a positive number of units corresponding to the length of the vector represented as a directed line segment. The magnitude of a vector A is denoted as $|A|$.

DEFINITION 2. The *scalar product* of a vector A by a scalar m, written simply as mA, is a vector having the direction of A and a magnitude that is the product of the magnitudes of m and A. If m is negative, the direction of A is reversed.

DEFINITION 3. To *add* two vectors A and B, we represent them as directed line segments to some scale at a common point O. Then, using the parallelogram construction, we form the sum as the diagonal element with A and B as sides. The magnitude of the sum is simply the length of the diagonal to the scale used.

EXERCISE A.1

Add vectors A and B (Fig. A.1) graphically to get $(A + B)$. Now add $(A + B)$ to C graphically. Show that the result is the same as adding B

189

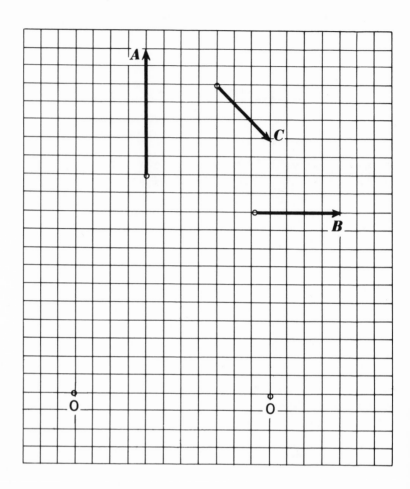

$$(A + B) + C \qquad\qquad (B + C) + A$$

FIGURE A.1

and C first and then adding this to A. Thus:

$$(A + B) + C = (B + C) + A$$

When an operation taken a number of times can be grouped in any manner, the operation is said to be *associative*. Thus, addition follows the associative law.

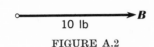

FIGURE A.2

EXERCISE A.2

Add vectors A and B (Fig. A.2) employing the parallelogram construction to sketch all vectors approximately to scale. Use the law of cosines for one of the triangles in the parallelogram to show that:

$$|A + B| = 7.37 \text{ lb}$$

Use the law of sines to show that $A + B$ is inclined by an angle α from the horizontal given as:

$$\alpha = 28.6°$$

EXERCISE A.3

Connect vectors A, B and C of exercise A.1, head to tail in any order in Fig. A.3. Now connect the exposed tail to the exposed head by an arrow going from the aforementioned tail to aforementioned head. Show that "closing the force polygon" in this way gives the sum of the vectors $A + B + C$. This process can be used for the addition of any number of vectors.

EXERCISE A.4

To subtract B from A in exercise A.2, add the vector $A + (-B)$. Show that:

$$|A - B| = 18.48 \text{ lb}$$

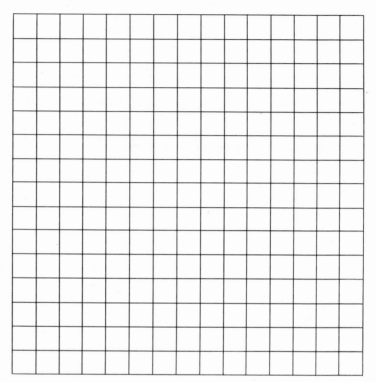

FIGURE A.3

Problems

▶ In the following problems, solve analytically unless instructed to do otherwise.

1. Add a 20-lb force pointing in the positive x direction to a 50-lb force at an angle of 45° to the x axis in the first quadrant and directed away from the origin.

2. Subtract the 20-lb force in the above problem from the 50-lb force.

3. Add the vectors shown in Fig. P.1 in the xy plane. Do this first analytically and then do it graphically using the force polygon.

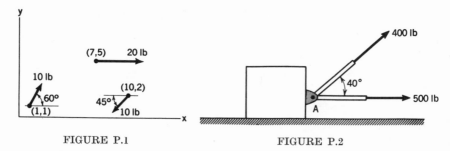

FIGURE P.1   FIGURE P.2

4. What is the sum of the forces transmitted by the structural members to the pin at A as shown in Fig. P.2?

5. Suppose in Prob. 4 we require that the total force transmitted by the members to pin A be inclined 12° to the horizontal. If we do not change the force transmitted by the horizontal member, what must be the new force for the other member? What is the total force?

6. Shown in Fig. P.3 is a simple slingshot about to be "fired." If the rubber band has a stretch of 3 lb per inch, what force does the band exert on the pellet and hand? The rubber band is 5 in. long when unstretched.

7. Forces are transmitted by two members in Fig. P.4 to pin A. If the sum of these forces is 700 lb directed vertically, what are the angles α and β?

8. In Fig. P.5, forces A (given as a horizontal 10-lb force) and B (vertical) add up to a force C that has a magnitude of 20lb. What is the magnitude of

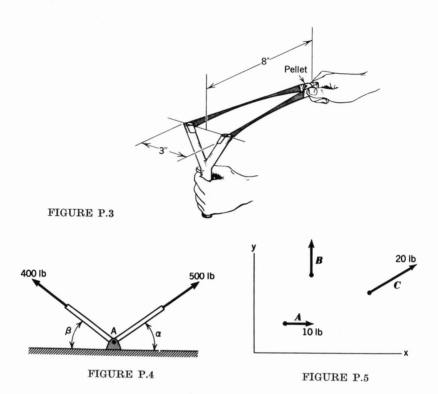

FIGURE P.3

FIGURE P.4   FIGURE P.5

force **B** and the direction of force **C**? (For the simplest results, use the force polygon, which for this case is a right triangle, and perform analytical computations.)

9. If the difference between forces **B** and **A** in Fig. P.5 is a force **D** having a magnitude of 25 lb, what is the magnitude of **B** and the direction of **D**?

DEFINITION 4. Consider a vector C and a pair of directions coplanar with C as shown in Fig. A.4. The process of finding two vectors C_1 and C_2 along directions 1 and 2 such that $C_1 + C_2 = C$ is called the resolution of vector C into two *component vectors*.

EXERCISE A.5

Find the vector components of C graphically in Fig. A.4. Show that:

$$|C_1| = 70\text{ lb} \qquad |C_2| = 56.6\text{ lb}$$

Verify the above results using trigonometric relations.

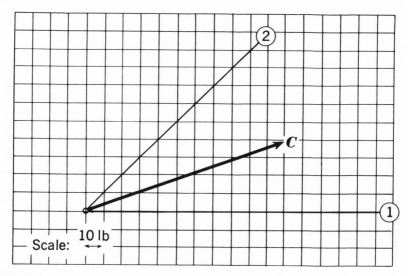

FIGURE A.4

DEFINITION 5. Consider a vector C and three orthogonal directions corresponding to coordinate axes x, y, and z (Fig. A.5). Sketch the resolution of C into a vector component C_3 along the z direction and a vector component C_4 in the xy plane. Now sketch the resolution of C_4 into the vector components C_1 and C_2 in the x and y directions, respectively. C_1, C_2, and C_3 are called the *orthogonal vector components* of C since they form a system of orthogonal vectors that add to give C.

The direction of C is given by the direction cosines as follows (see Fig. A.6):

$$\cos (C, x) = \cos \alpha = l$$
$$\cos (C, y) = \cos \beta = m$$
$$\cos (C, z) = \cos \gamma = n$$

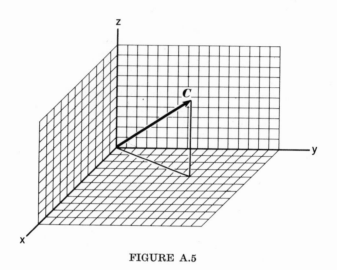

FIGURE A.5

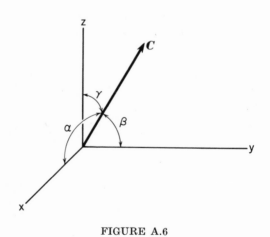

FIGURE A.6

Show with the aid of your sketch on p. 196 that:

$$|C_3| = |C|n$$

Explain how, by proceeding differently in Fig. A.5, you could show that:

$$|C_1| = |C|l$$

and that $|C_2| = |C|m$. We shall denote $|C|l$ as C_x, $|C|m$ as C_y, and $|C|n$ as C_z. Note that C_x, C_y, and C_z are scalar quantities.

Definition 6. The *rectangular component* of a vector C in some direction s (see Fig. A.7) is simply the component along s for a set of orthogonal components (not completely shown) with s as one of the orthogonal directions. Hence, $C_s = |C| \cos \delta$.

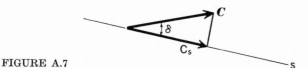

FIGURE A.7

Problems

10. Resolve the 20-lb force in problem 1 into two component forces inclined at 45° above and below the *x* axis.

11. Resolve the 1000-lb force in Fig. P.6 into a set of component vectors along the directions shown (in solid lines).

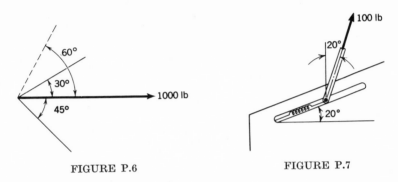

FIGURE P.6   FIGURE P.7

12. Resolve the 100-lb force shown in Fig. P.7 into a set of components along the slot shown and in the vertical direction.

13. Resolve the force *F* in Fig. P.8 into a component perpendicular to *AB* and a component parallel to *BC*.

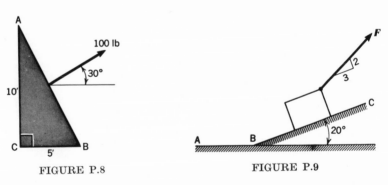

FIGURE P.8   FIGURE P.9

14. Resolve the 100-lb force in Fig. P.9 into components parallel to *BC* and *AB*.

15. A 1000-lb force is resolved into components along AB and AC as shown in Fig. P.10. If the component along AB is 700 lb, determine the angle α and the value of the component along AC.

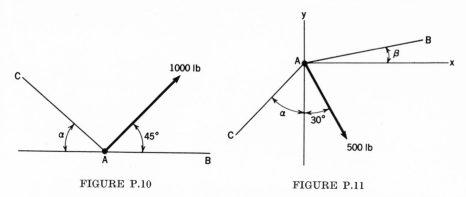

FIGURE P.10   FIGURE P.11

16. The 500-lb force shown in Fig. P.11 is to be resolved into components along the AB and AC directions measured by the angles α and β. If the component along AC is to be 800 lb and the component along AB is to be 1000 lb, compute α and β.

17. What is the orthogonal component of the 1000-lb force in Fig. P.6 in the direction (dotted line) inclined at 60° from the force?

18. The orthogonal components of a force are:

x component 10 lb in positive x direction
y component 20 lb in positive y direction
z component 30 lb in negative z direction

(a) What is the magnitude of the force itself?
(b) What are the direction cosines of the force?

19. Using results from Definition 5, show that:

$$l^2 + m^2 + n^2 = 1$$

20. A force vector of magnitude 100 lb has a line of action with direction cosines $l = 0.7$, $m = -0.2$ relative to some reference xyz. If the vector points in a direction away from the origin, what is the vector representation in terms of orthogonal components? (Use the positive value for n in this problem.)

21. What are the components of the 1000-lb force shown in Fig. P.12 along the indicated coordinate directions?

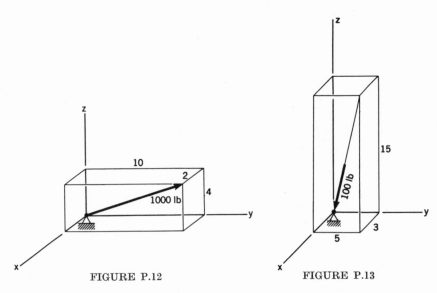

FIGURE P.12   FIGURE P.13

22. What are the force components of the 100-lb force shown in Fig. P.13? What are the direction cosines associated with the 100-lb force?

23. What is the orthogonal force component in the x direction transmitted to pin A by the three members shown in Fig. P.14?

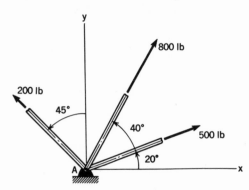

FIGURE P.14

24. What are the rectangular components of the 100-lb force shown in Fig. P.15? What are the direction cosines for this force?

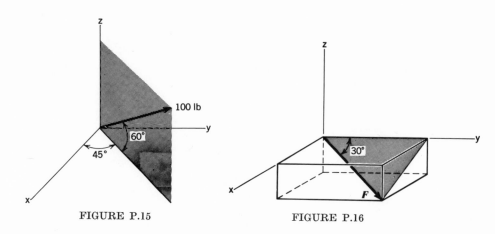

FIGURE P.15   FIGURE P.16

25. The x and z components of the force F shown in Fig. P.16 are known to be 100 lb and -30 lb, respectively. What is the force F and what are its direction cosines?

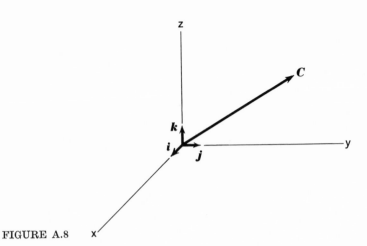

FIGURE A.8

DEFINITION 7. A *unit vector* is one having no dimensions and a unit magnitude. Unit vectors in the coordinate directions of cartesian reference axes x, y, and z are denoted as i, j, and k, respectively (see Fig. A.8). We now can see that any vector C can be given as follows:

$$C = C_x i + C_y j + C_z k$$

Explain then how from the vector equation

$$F = ma$$

we may arrive at the following three scalar equations:

$$F_x = ma_x \qquad F_y = ma_y \qquad F_z = ma_z$$

Another way of forming a unit vector in any direction (say the direction of vector C) is as follows:

$$\hat{c} = \frac{C}{|C|}$$

Explain why $\hat{c}$ qualifies as a unit vector in the direction of C.

EXERCISE A.6

Given vector $C = 10i + 16j - 8k$, show that the direction cosines of this vector are: $l = .488$, $m = .782$, $n = -.390$. Also show that $\hat{c} = .488i + .782j - .390k$.

EXERCISE A.7

For the vectors shown in Fig. A.9, demonstrate that:

$$A + B + C + D + E = 60i + 0j + 30k$$

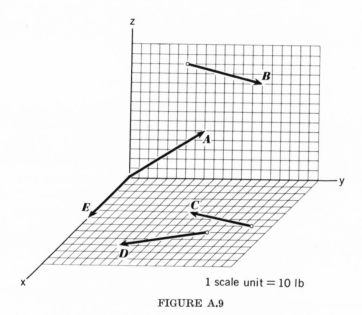

1 scale unit = 10 lb

FIGURE A.9

DEFINITION 8. The dot product (or scalar product) of two vectors A and B is denoted $A \cdot B$ and is a scalar given as:

$$A \cdot B = |A||B| \cos \alpha$$

where α is the smaller angle between the vectors (see Fig. A.10). We can say that $A \cdot B$ equals the rectangular projection of A along B times the magnitude of B, or the projection of B along A times the magnitude

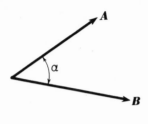

of A. The result is positive if the projected vector and the second vector are in the same direction—otherwise the result is negative. Show that:

1. $(mA) \cdot (nB) = mnA \cdot B$
2. $A \cdot B = B \cdot A$ (the dot product is commutative)
3. $A \cdot B = A_x B_x + A_y B_y + A_z B_z$
4. $A \cdot A = A^2$

FIGURE A.10

EXERCISE A.8

Find $A \cdot B$ in Fig. A.11 using the definition of the dot product directly. Then compute $A \cdot B$ using rectangular components of the vector.

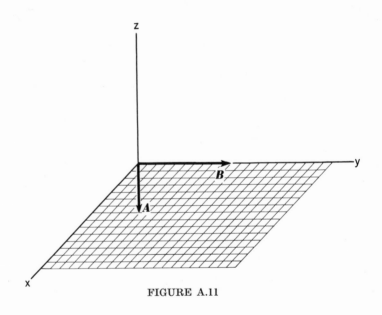

FIGURE A.11

Problems

26. What is the sum of the following set of three vectors?

$A = 6i + 10j + 16k$

$B = 2i - 3j$

C is a vector in the xy plane at an inclination of 45° to the positive x axis and directed away from the origin. It has a magnitude of 25 lb.

27. Subtract vector C from the sum of vectors A and B in the above problem.

28. A vector A has a line of action that goes through the coordinates $(0, 2, 3)$ and $(-1, 2, 4)$. If the magnitude of this vector is 10 units, express the vector in terms of the unit vectors i, j, and k.

29. Express the force shown in Fig. P.17 in terms of the unit vectors i, j, and k.

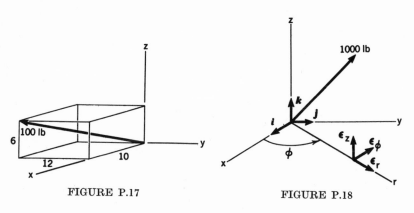

FIGURE P.17 FIGURE P.18

30. The displacement from the origin to some point xyz is denoted as r and is called the *position vector*. What is the corresponding unit vector $\hat{r}$ for $x = 3$, $y = 4$, and $z = -3$? Express the result in terms of rectangular components.

31. Curvilinear coordinate systems also have associated with them sets of unit vectors. These do not have fixed directions in space as do i, j, and k (see Fig. P.18). Express the unit vector j in terms of unit vectors ϵ_r, ϵ_ϕ, and

205

ϵ_z. Express the 1000-lb force going through the origin and through point (2, 4, 4) in terms of the unit vectors i, j, k and $\epsilon_r, \epsilon_\phi, \epsilon_z$ with $\phi = 45°$.

32. Show that:

$$\cos (A, B) = ll' + mm' + nn'$$

where l, m, n, and l', m', n' are direction cosines of A and B, respectively, with respect to the given xyz reference.

33. What is the component of the force vector in Prob. 20 along a direction having the direction cosines $l = -0.3$, $m = 0.1$, $n = 0.95$ for the xyz reference?

34. Given the following vectors:

$$A = 10i + 20j + 3k$$
$$B = -10j + 12k$$

What is $A \cdot B$? What is $\cos (A, B)$? What is the projection of A along B?

35. Given the vectors:

$$A = 16i + 3j, \qquad B = 10k - 6i, \qquad C = 4j$$

Compute

(a) $C(A \cdot C) + B$

(b) $-C + [B \cdot (-A)]C$

36. What is the dot product of the force vector $(10i + 6j - 3k)$ lb and the displacement vector $(6i - 2j)$ ft?

37. A constant force given as $2i + 3k$ moves a particle along a straight line from position $x = 10$, $y = 20$, $z = 0$ to position $x = 3$, $y = 0$, $z = -10$. If the coordinates of the xyz reference are given in feet units, how much work does the force do?

38. Explain why the following operations are meaningless:

(a) $(A \cdot B) \cdot C$

(b) $(A \cdot B) + C$

39. Shown in Fig. P.19 is a rectangular parallelepiped. (a) Find the dot product of the vectors represented by the diagonals from A to F and from

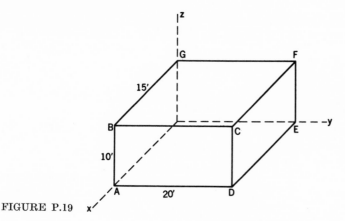

FIGURE P.19

D to G. What is the angle between them? (b) Find the dot product of the diagonals from G to C and from G to D. What is the angle between them?

40. What is the rectangular component of the 500-lb force shown in Fig. P.20 along the diagonal from B to A?

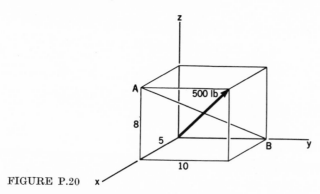

FIGURE P.20

41. An electrostatic field E exerts a force on a charged particle of qE where q is the charge of the particle. If we have for E:

$$E = 6i + 3j + 2k \text{ dynes/coulomb}$$

what work is done by the field if a particle with a unit charge moves along a straight line from the origin to position $x = 2$ cm, $y = 4$ cm, $z = -4$ cm?

42. Given the following force expressed as a function of position:

$$F = (10x - 6)i + x^2zj + xyk$$

What are the direction cosines of the force at position $(1, 2, 2)$? What is the position along the x coordinate where $F_x = 0$? Plot F_y versus the x coordinate for an elevation $z = 1$.

43. Given the force $\boldsymbol{F}$ as a function of position in the following form:

$$\boldsymbol{F} = (y^2 + 2z)\boldsymbol{i} + (10 + 3y^2)\boldsymbol{j} + (z^3 + xy)\boldsymbol{k}$$

How much work is done as the force moves a particle along a straight line from $(0, 2, 4)$ to $(-3, 5, -7)$? Would the result change if another path were chosen but with the same endpoints?

*44. Given the following force expressed as a function of position in space as

$$\boldsymbol{F} = xy\boldsymbol{i} + z^2\boldsymbol{j} + 10\boldsymbol{k}$$

What work does this force do if it goes along a straight line from position $(2, 4, 4)$ to position $(3, 9, -12)$?

DEFINITION 9. The cross product between two vectors A and B, denoted as $A \times B$, is a vector C having the magnitude

$$|C| = |A||B| \sin \alpha$$

where α is the smaller angle between A and B (see Fig. A.12). The direction of C is normal to the plane of A and B when these vectors are moved to intersect, with a sense corresponding to the right-hand rule as you rotate from the first vector A in the statement to the second vector B through the angle α.

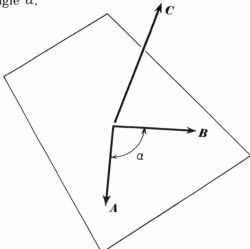

FIGURE **A.12**

1. Explain why $A \times B = -B \times A$.
2. Noting that:

$$i \times j = k$$
$$k \times j = i$$

express all possible cross products of unit vectors i, j, and k.
3. Show that:

$$A \times B = (A_y B_z - A_z B_y)i + (A_z B_x - A_x B_z)j$$
$$+ (A_x B_y - A_y B_x)k$$

4. Show that the above result can be given by formally carrying out the following determinant:*

$$\begin{vmatrix} A_x & A_y & A_z \\ B_x & B_y & B_z \\ i & j & k \end{vmatrix}$$

EXERCISE A.9

Using Fig. A.11, find the cross product $A \times B$ by using the definition directly and then by using rectangular components.

---

*To carry out a 3×3 determinant, repeat the first two rows below the determinant. Then form diagonals as shown below. Multiply along the diagonal, remembering to include -1 in the product of each of the dashed diagonals. The sum of these six expressions is then the determinant.

$$\begin{vmatrix} a & b & c \\ d & e & f \\ g & h & i \end{vmatrix} =$$

Problems

45. If $A = 10i + 6j - 3k$ and $B = 6i$, find $A \times B$ and $B \times A$. What is the magnitude of the resulting vector? What are its direction cosines relative to the xyz reference in which A and B are expressed?

46. The force on a charge moving through a magnetic field B is given as:

$$F = qV \times B$$

where q is the magnitude of the charge in coulombs,
F is the force on the body in newtons,
V is the velocity vector of the particle in meters per second, and
B is the magnetic flux density in webers per meter^2.

Suppose an electron moves through a uniform magnetic field of 100 webers/cm^2 in a direction inclined 30° to the field, as shown in Fig. P.21, with a speed of 100 meters/sec. What are the force components on the electron in the horizontal and vertical directions? The charge of the electron is 1.6018×10^{-19} coulomb.

47. (a) If $A \cdot B = A \cdot B'$, does B necessarily equal B'? Explain.
(b) If $A \times B = A \times B'$, does B necessarily equal B'? Explain.

48. If vectors A and B in the xy plane have a dot product of 50 units and if the magnitudes of these vectors are 10 units and 8 units, respectively, what is $A \times B$?

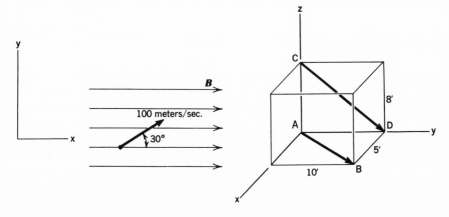

FIGURE P.21   FIGURE P.22

211

49. What are the cross and dot products for the vectors **A** and **B** given as:

$$A = 6i + 3j + 4k$$
$$B = 8i - 3j + 2k$$

50. What is the cross product of the displacement vector from A to B in Fig. P.22 times the vector from C to D?

51. Figure P.23 shows a flow pattern. The velocity of a particle of the flow is given as:

$$V = 10i + 16j + 2k \text{ ft/sec}$$

What is the cross product of the position vector **r** given as:

$$r = 3i + 2j + 10k \text{ ft}$$

times the vector **V**? Give the proper units.

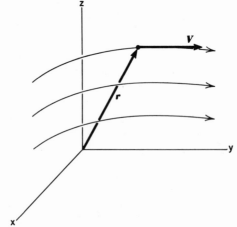

FIGURE P.23

52. Making use of the cross product and the edges of ABC, give the unit vector **n** normal to the inclined surface ABC in Fig. P.24.

53. A pyramid is shown in Fig. P.25. If the height of the pyramid is 300 ft find the unit vectors n_1 and n_2 normal to the faces ADB and BDC. What is the angle between planes ADB and DBC?

54. In Fig. P.26 is shown an inclined pyramid. If the coordinates of vertex E are (5, 50, 80) ft, what is the angle between faces ADE and BCE?

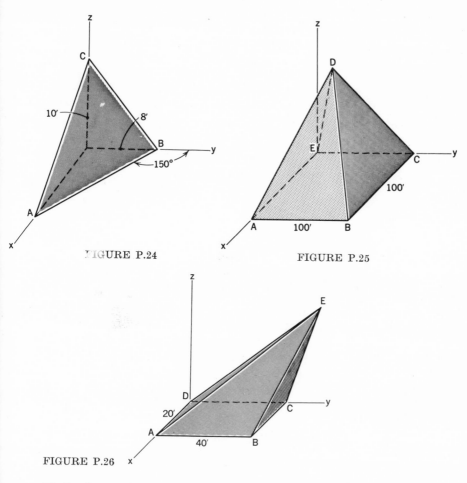

FIGURE P.24

FIGURE P.25

FIGURE P.26

55. In the previous problem what is the area of each face of the pyramid? What is the projected area of faces ADE and ABE along the direction ϵ where:

$$\epsilon = 0.6i - 0.8j$$

DEFINITION 10. The *scalar triple product* for a set of vectors A, B, and C is defined as:

$$(A \times B) \cdot C$$

Note next that the volume of a parallelepiped (Fig. A.13) equals the area of the base *abcd* times the slant height h, which is the perpendicular distance between the base *abcd* and the opposite parallel face *efgh*.

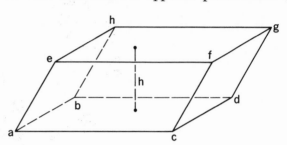

FIGURE A.13

Now explain why $(A \times B) \cdot C$ is the volume of a parallelepiped formed from vectors A, B, and C positioned to be concurrent. From this explain why the following relations hold:

$$(A \times B) \cdot C = -(A \times C) \cdot B = -(C \times B) \cdot A$$

EXERCISE A.10

Carry out the triple scalar product $(A \times B) \cdot C$ using rectangular components and then compare this result with the following determinant:

$$\begin{vmatrix} A_x & A_y & A_z \\ B_x & B_y & B_z \\ C_x & C_y & C_z \end{vmatrix}$$

Problems

56. Given the following vectors:

$$A = 10i + 6j$$
$$B = 3i + 5j + 10k$$
$$C = i + j - 3k$$

Find

    (a) $(A + B) \times C$

    (b) $(A \times B) \cdot C$

    (c) $A \cdot (B \times C)$

57. Shown in Fig. P.27 is a parallelepiped. The surface *abcd* is in the *xz* plane. Compute the volume using vector analysis.

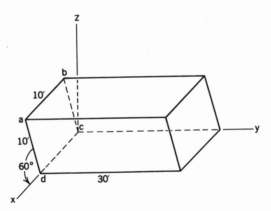

FIGURE P.27

58. What is the component of the cross product $A \times B$ along the direction *n* where:

$$A = 10i + 16j + 3k$$
$$B = 5i - 2j + 2k$$
$$n = 0.8i + 0.6k$$

59. Using the scalar triple product find the area projected onto the plane N from the surface ABC (see Fig. P.28). Plane N is infinite and is normal to the vector:

$$r = 50i + 40j + 30k \text{ ft}$$

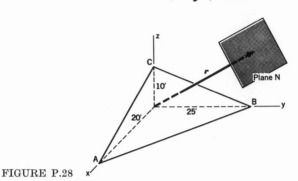

FIGURE P.28

60. We can always find a reference so that the components of any three vectors A, B, and C can be expressed as:

$$A = A_x i + A_y j + A_z k, \ B = B_x i + B_y j, \ C = C_x i$$

Explain how you can select such a reference.

61. Using vector components in the preceding problem, show that for any three vectors A, B, and C we can say:

$$A \times (B \times C) = B(A \cdot C) - C(A \cdot B)$$

62. Given the following vectors:

$$A = 6i + 10j - 10k$$
$$B = 3j - 10k$$
$$C = 3i + 15j$$

Compute the vector triple product $A \times (B \times C)$.

63. Given the following vectors:

$$A = 10i + 6j$$
$$B = 3i - 5j + 10k$$
$$C = i + j - 3k$$

Find

(a) $(A + B) \times C$

(b) $(A \times B) \times C$

(c) $A \times (B \times C)$

(d) $A \cdot (B \times C)$

64. Using the vectors of Prob. 62 compute the following:

$$[(A + B) \cdot C]A \times (B \times C)$$

ANSWERS TO SELECTED PROBLEMS

CHAPTER 1

1 (a) $\dfrac{(M)}{(L^3)}$

   (b) $1\dfrac{\text{gram}}{(\text{cm})^3} \equiv 1.94\dfrac{\text{slugs}}{(\text{ft})^3}$ or $1\dfrac{\text{slug}}{(\text{ft})^3} \equiv .515\dfrac{\text{gram}}{(\text{cm})^3}$

      $1\dfrac{\text{gram}}{(\text{cm})^3} \equiv 6.25\dfrac{\text{lbm}}{(\text{ft})^3}$ or $1\dfrac{\text{lbm}}{(\text{ft})^3} \equiv 0.0162\dfrac{\text{gram}}{(\text{cm})^3}$

2 (a) $\dfrac{L}{t^2}$

   (b) $\dfrac{F}{L^2} \equiv \dfrac{M}{Lt^2}$

   (c) 1

   (d) $\dfrac{L^3}{M}$

3 25,000 miles/hour

   5360 miles/hour

5 (a) $\left(\dfrac{M}{LT}\right)$

   (b) $1\dfrac{\text{slug}}{\text{ft sec}} \equiv 480\dfrac{\text{gram}}{\text{cm sec}} = 480 \text{ poise}$

6 dimensionless

8 $\dfrac{M}{t^2}$

19 1.228 lb

20 12.29 ft/sec^2

21 0.0120

22 197,200 miles

24 6.75 × 10^{21} tons mass

25 $F_e = 2.8 \times 10^{-8}$ Newton
$F_g = 5.52 \times 10^{-51}$ Newton

CHAPTER 2

1 $3i + 4j + 5k$
$l = 0.425; m = 0.566; n = 0.706$

2 $4i - 16j - 3k$

3 $7i - 4j - 11k$

4 15.16 ft; $l = -0.461; m = -0.660; n = 0.593$

5 $r = xi \pm \sqrt{1 - x^2}j$

6 $r = \pm 2\sqrt{2z}j + zk$

7 $r = 6i + 10.16j - 2.4k$

8 149,000 lb-ft

9 -600 lb-ft

10 2820 lb-ft

11 About A: $-8940k$ lb-ft
About B: $-4475k$ lb-ft
About C: 0

12 $M = -42i + 100j + 30k$

13 $M = -84i + 94j - 46k$

14 $M = 11.86i + 9j + 23.2k$

15 $M_A = \dfrac{10}{\sqrt{3}}ak - \dfrac{10}{\sqrt{3}}aj$ lb-ft

$M_B = \dfrac{10}{\sqrt{3}}ai - \dfrac{10}{\sqrt{3}}aj$ lb-ft

$M_C = \dfrac{10}{\sqrt{3}}ai - \dfrac{10}{\sqrt{3}}ak$ lb-ft

16 $M_1 = -450k$ lb-ft; $M_2 = -1200k$ lb-ft

17 $M_A = 163.2i - 244.8j + 81.6k$ lb-ft
$M_B = 163.2i - 652.8j - 326.4k$ lb-ft

18 $M_A = -2160i + 850j - 6500k$ lb-ft
$M_B = -11220i - 4330j - 4330k$ lb-ft

19 $M_E = (7.29F_{AB} + 8.59F_{CD} - 10,000)\mathbf{i} + (2.92F_{AB} - 5.15F_{CD})\mathbf{j}$
$+ (-2.92F_{AB} + 5.15F_{CD})\mathbf{k}$

20 $M = 1.33F$

21 (a) $M_x = -42; M_y = 100; M_z = 30$
(b) 7.79
(c) -6.6

22 $F = 146.3$ lb; $M = 77.2$ lb-ft

23 $M_{AC} = 850$ lb-ft; $M_{BC} = -11,220$ lb-ft

24 -5.22 lb-ft

25 83.7 lb-ft

26 8.4 lb-ft

27 76.5 lb-ft

28 5770 lb-ft

29 $\mathbf{M} = -36\mathbf{i} + 41\mathbf{j} + 38\mathbf{k}$
$l = -0.541; m = 0.616; n = 0.572$

30 5.52 units

31 (a) $-50\mathbf{i}$
(b) $-50\mathbf{i}$
(c) 0
(d) 0

32 (a) $\mathbf{M} = -21.21\mathbf{j} - 21.21\mathbf{k}$ lb-ft
(b) $M_{AD} = -24.4$

33 1150 lb-ft

34 $\mathbf{M}_A = -2250\mathbf{k}$ lb-ft

35 $\mathbf{M}_A = -7270\mathbf{k}$ lb-ft; $M_B = 21,010\mathbf{k}$ lb-ft

36 $\mathbf{M}_T = 1000\mathbf{i} - 1500\mathbf{j} - 1000\mathbf{k}$ lb-ft. Moment about (3,4,2) is $\mathbf{M}_T$ since couples are free vectors. $M_r = -930$ lb-ft. The total force is (0) since a couple is composed of two equal and opposite forces.

37 (a) $C = 35$ lb-ft; $\theta = 36.9°$
(b) $d = 7$ ft

38 $\mathbf{C} = 35.4\mathbf{i} + 22.3\mathbf{j} + 80\mathbf{k}$

39 $\mathbf{C}_T = -450\mathbf{j} + 250\mathbf{k}$

41 65.3 lb-ft; 36.5 lb-ft

42 1039 lb-ft

43 $100\mathbf{i} + 50\mathbf{j} - 7.79\mathbf{k}$ lb-ft

44 -470 lb-ft; $l = 0; n = \pm0.8$

47 $\mathbf{F}_A = -866\mathbf{j} + 500\mathbf{k}$ lb
$\mathbf{M}_A = 4330\mathbf{i}$ ft-lb
$\mathbf{F}_B = -866\mathbf{j} + 500\mathbf{k}$ lb
$\mathbf{M}_B = 2330\mathbf{i}$ ft-lb

48 At A: $\mathbf{F} = -354\mathbf{j} - 354\mathbf{k}$
$\mathbf{C}_A = -3540\mathbf{i}$ ft-lb

At B: $\mathbf{F} = -354\mathbf{j} - 354\mathbf{k}$
$\mathbf{C}_B = 7080\mathbf{i}$ ft-lb

Add $\mathbf{C} = -100\mathbf{i}$

At A: $\mathbf{F} = -354\mathbf{j} - 354\mathbf{k}$
$\mathbf{C} = -3640\mathbf{i}$

At B: $\mathbf{F} = -354\mathbf{j} - 354\mathbf{k}$
$\mathbf{C} = 6980\mathbf{i}$

49 (a) $\mathbf{F} = 5\mathbf{j}; \mathbf{C} = 5\mathbf{i}$
(b) $\mathbf{F} = 5\mathbf{j}; \mathbf{C} = 0$
(c) $\mathbf{F} = 5\mathbf{j}; \mathbf{C} = -5\mathbf{k}$
(d) $\mathbf{F} = 5\mathbf{j}; \mathbf{C} = 5\mathbf{i} - 5\mathbf{k}$

50 $\mathbf{F} = 3\mathbf{i} - 6\mathbf{j} + 4\mathbf{k}$ lb; $\mathbf{C} = -16\mathbf{i} - 40\mathbf{j} - 48\mathbf{k}$

51 $\mathbf{F} = 3\mathbf{i} + 10\mathbf{j} + 16\mathbf{k}$ lb; $\mathbf{C} = -120\mathbf{i} + 44\mathbf{j} - 5\mathbf{k}$ lb-ft

52 $\mathbf{F} = 6\mathbf{i} + 3\mathbf{j} + 10\mathbf{k}$ lb; $\mathbf{C} = 46\mathbf{i} - 42\mathbf{j} - 15\mathbf{k}$

53 $\mathbf{F} = 10\mathbf{i} + 3\mathbf{j} - 2\mathbf{k}; \mathbf{C} = 13\mathbf{i} - 22\mathbf{j} + 32\mathbf{k}$

54 $\mathbf{F} = -353\mathbf{i} - 353\mathbf{k}$ lb; $\mathbf{C} = -1000\mathbf{i} - 3530\mathbf{j}$ lb-ft

55 $\mathbf{F}_R = 9\mathbf{i} + 10\mathbf{j} + 9\mathbf{k}$ lb; $\mathbf{C}_R = 6\mathbf{i} + 10\mathbf{j} + 6\mathbf{k}$ lb-ft

56 $\mathbf{F} = -480\mathbf{i} - 685\mathbf{j} + 549\mathbf{k}; \mathbf{C} = 1088\mathbf{i} - 9330\mathbf{j} - 10690\mathbf{k}$

57 $\mathbf{C}_T = 0; \mathbf{F}_T = 24{,}000\mathbf{j} - 5000\mathbf{k}$ lb

58 $\mathbf{F}$ must intercept the x axis at $x = 0.834$

59 $y = 5$ ft

60 0.576 ft from top

61 1 ft from A

62 $\mathbf{F}_R = 0; \mathbf{C}_R = -7\mathbf{i} + 3\mathbf{j}$

63 (a) $\mathbf{F}_R = 35.3\mathbf{i} + 100\mathbf{j} + 35.3\mathbf{k}; \mathbf{C}_R = -247\mathbf{i} + 147\mathbf{k}$
(b) $\mathbf{F}_R = 35.3\mathbf{i} + 100\mathbf{j} + 35.3\mathbf{k}; \mathbf{C}_R = -100\mathbf{i} + 500\mathbf{k} - 176.7\mathbf{j}$

64 At A: $\mathbf{F}_R = -204\mathbf{i} + 604\mathbf{j} + 408\mathbf{k}$
$\mathbf{C}_R = 8160\mathbf{i} + 2032\mathbf{j} + 3264\mathbf{k}$

At B: $\mathbf{F}_R = -204\mathbf{i} + 604\mathbf{j} + 408\mathbf{k}$
$\mathbf{C}_R = 2032\mathbf{j} - 816\mathbf{k}$

65 $\mathbf{F}_R = 54.8\mathbf{i} - 46.5\mathbf{j} - 310\mathbf{k}$ lb
$\mathbf{C}_R = -211.5\mathbf{i} - 317.8\mathbf{j} - 50\mathbf{k}$ ft-lb

66 43.3 lb; -1110 ft-lb

67 $F_1 = 26.9$ lb; $F_2 = 17.6$ lb; $F_3 = 35.5$ lb

68 $\mathbf{F}_R = -31.8\mathbf{j} - 238\mathbf{k}$ lb; $\mathbf{C}_R = -1297\mathbf{i} - 690\mathbf{j} + 775\mathbf{k}$

69 $\mathbf{F} = 74.6\mathbf{j} + 49.2\mathbf{i}$ lb; $\bar{x} = 0.67$

70 $\mathbf{F}_R = 4.3\mathbf{i} - 4.65\mathbf{j}$ lb; $\bar{x} = -26.4$ ft

71 $\mathbf{F}_R = 2100\mathbf{i} - 500\mathbf{j}$ lb; $\bar{x} = -0.05$ ft

72 $\mathbf{F}_R = -250\mathbf{i} - 2300\mathbf{j}$ lb; $\bar{x} = 15.66$ ft

73 $\mathbf{F}_R = 100\mathbf{i} - 75\mathbf{j}$ lb; $\bar{x} = 29.7$ ft

74 $\mathbf{F}_R = 326.8\mathbf{j} - 200\mathbf{k}$; $\bar{y} = 40.5$ ft
   At A: $\mathbf{F}_R = 326.8\mathbf{j} - 200\mathbf{k}$ lb
           $\mathbf{C}_R = -8100\mathbf{i}$ ft-lb

75 $\mathbf{F}_R = -90\mathbf{j}$ lb; $\bar{x} = 21.2$ ft

76 (a) $\mathbf{F}_R = -100\mathbf{k}$; $\bar{x} = 2.5$; $\bar{y} = 2.2$
   (b) $\mathbf{C}_R = 280\mathbf{i} + 450\mathbf{j}$

77 $\mathbf{C} = -600\mathbf{j} - 300\mathbf{k}$ ft-lb

78 $\mathbf{F}_R = -120\mathbf{j}$ lb; $\bar{z} = 0.471$ ft; $\bar{x} = 6.97$ ft

79 $y' = 7.5$ ft; $x' = -12.5$ ft

80 $\mathbf{C}_R = 1000\mathbf{i}$ lb-ft

81 $\mathbf{F}_R = 200\mathbf{j}$ lb; $\bar{z} = 0.9$ ft; $\bar{x} = 5$ ft

82 $\mathbf{F} = 86.6\mathbf{i} + 1050\mathbf{j}$ lb
   $\mathbf{M}_{CG} = 2100\mathbf{i} - 173.2\mathbf{j} + 2770\mathbf{k}$ ft-lb; 2770 ft-lb

83 $\mathbf{F}_R = -1150\mathbf{j}$ lb; $\bar{x} = 12.58$ ft

84 (a) $\mathbf{F} = 35\mathbf{i} + 158\mathbf{j} + 15\mathbf{k}$ lb
   (b) $30\mathbf{i} + 158\mathbf{j}$ lb

85 $\mathbf{F} = 3166\mathbf{k}$ lb

86 $\mathbf{F} = 51{,}200\mathbf{k}$ oz

87 $\bar{x} = 10.4$ in.; $\bar{y} = \bar{z} = 0$

88 $\bar{x} = 16.1$ in.

89 (a) $\gamma = 200 - 5y + 6.25z$ lbf/ft^3
   (b) $\bar{y} = 4.78$ ft; $\bar{x} = 2$ ft

90 $\bar{x} = 2.095$ ft; $\bar{y} = 5.15$ ft

91 $\bar{z} = \dfrac{(\gamma' + 4\gamma_0)}{2(\gamma' + 9\gamma_0)}\, h$
   $\bar{x} = \bar{y} = 0$

92 $\bar{z} = \dfrac{\gamma_0^2 h^2\,(\gamma' + 4\gamma_0) - 15a^2 b^2\left[2\gamma_0 + (\gamma' - \gamma_0)\dfrac{b^2}{h^2}\right]}{2\gamma_0^2 h^2\,(\gamma' + 9\gamma_0) - 20a^2 b\left[3\gamma_0 + (\gamma' - \gamma_0)\dfrac{b^2}{h^2}\right]}$

93 $a/3$ from the bottom.

94 $\bar{x} = 1.2$ ft; $\bar{y} = 1.125$ ft

95 $\bar{x} = 2.25$ ft, $\bar{y} = 2.4$ ft

96 $\mathbf{F}_R = -p_0\,\dfrac{ab}{2}\,\mathbf{k}$; $\bar{x} = \dfrac{b}{2}$, $y = \dfrac{2}{3}\,a$

97 $\mathbf{F}_R = 2{,}310{,}000$; $\bar{y} = 12.53$ ft, $\bar{x} = 20$ ft

98 $(3,2,4)$; $(1.33,2,0)$; $\mathbf{C}_R = -192{,}800\mathbf{j}$ lb-ft

99 $F = -748$; $\bar{s} = 2.23$ ft

100 $\mathbf{F}_R = 3000\mathbf{k}$

101 $F_R = 18{,}725{,}000$; $\bar{x} = 14.6$ ft, $\bar{y} = 5$ ft

102 $F_R = 262$ lb

103 $F_R = 2360$ lb

104 13,200 lb

105 $(F_x)_R = 406.5$ lb; $(M_D)_T = -17,558k$ lb-ft

106 200 lb, passing 10 ft to the left of beam.

107 $F_R = -6330j$ lb; $\bar{x} = 14.5$ ft

108 (a) $F = -\dfrac{20}{\pi}\, j$ acts at center of beam

    (b) $F = -\dfrac{20}{\pi}\, j; \; C = \dfrac{100}{\pi}\, k$

109 $F_R = -1549j$ lb   acting at $x = 20.1$ ft

110 Simplest resultant:

    $F_R = 160j$

    At $\bar{x} = 30.8$ ft

    $F_A = 160j$ lb; $M_A = 4940k$ ft-lb

111 $\bar{x} = 19.4$ ft; $\bar{y} = 8.13$ ft

112 $\bar{x} = 21.2$ ft; $\bar{y} = 4.54$ ft

CHAPTER 3

14 $T_{BA} = 37.8$ lb; $T_{BC} = 26.8$ lb

15 $T_{BC} = 113$ lb

16 $C_1 = C_2 = 671$ lb

    $A_y = 1000$ lb; $A_x = 447$ lb

    $B_y = 1000$ lb; $B_x = 447$ lb

17 $A_x = -17.5$ tons; $A_y = -9$ tons

    $B_x = 17.5$ tons; $B_y = 19$ tons

    $C_x = -17.5$ tons; $C_y = 21$ tons

18 $W = 270$ lb

19 $F = \dfrac{W}{2}\left(\dfrac{r_2 - r_1}{r_2}\right)$

20 $T = 30$ ft-tons

21 $W_{\max} = 2.5$ tons

    $B_y = 10.8$ tons; $A_y = 1.7$ tons

    $C_y = 1.7$ tons; $M_C = 34$ ft-tons

    $D_y = 1.7$ tons; $M_D = 85$ ton-ft

22 36.6° with vertical

23 3.1 ft

24 300 lb

25 18.75 ton-ft

26 $B = 735$ lb; $A_y = 925$ lb; $A_x = 0$

27 $A_x = 0$; $A_y = 700$ lb; $M = 8670$ lb-ft

28 43 lb at right end
87 lb at pin
237 lb at left end
-2430 lb-ft at left end

29 $B_y = 27.8$ lb
$A_y = 5.5$ lb
1005.5 lb at left support
10,099 lb-ft at left support
293.8 lb at right support
2766 lb-ft at right support

30 $\mathbf{A} = -59\mathbf{i} + 269\mathbf{j}$ lb

31 $\mathbf{C} = 5015\mathbf{i} + 2500\mathbf{j}$ lb

32 $\mathbf{A} = -337\mathbf{i} + 244\mathbf{j}$; $\mathbf{C} = 337\mathbf{i} + 206\mathbf{j}$

33 $T_D = 103.5$ lb; $T_C = 32.5$ lb; $T_A = 78$ lb

34 $T_C = 34.2$ lb; $T_D = 11.31$ lb; $T_A = 61.7$ lb

35 $\mathbf{F}_A = 408\mathbf{i} - 408\mathbf{j} - 818\mathbf{k}$ lb
$\mathbf{M}_A = -8660\mathbf{i} - 1635\mathbf{j} + 12,730\mathbf{k}$ lb-ft
$\mathbf{F}_B = 408\mathbf{i} - 408\mathbf{j} - 818\mathbf{k}$ lb
$\mathbf{M}_B = -8660\mathbf{i} + 8660\mathbf{j} - 8660\mathbf{k}$ lb-ft

36 $F = 57.2$ lb; $A_y = 160$ lb; $B_y = 93.2$ lb
$B_z = 22.6$ lb; $A_z = 17.8$ lb

37 $\mathbf{F}_A = -500\mathbf{j} + 1500\mathbf{k}$ lb; $\mathbf{M}_A = 26,200\mathbf{i} - 9100\mathbf{k}$
$\mathbf{F}_B = -500\mathbf{j}$ lb; $\mathbf{M}_B = -9100\mathbf{k}$ lb-ft

38 $\mathbf{F}_A = -10\mathbf{i} - 3\mathbf{j} + 500\mathbf{k}$
$\mathbf{M}_A = -560\mathbf{i} - 4300\mathbf{j} + 10\mathbf{k}$ lb-ft

39 $\mathbf{F}_A = -10\mathbf{i} - 3\mathbf{j} + 2500\mathbf{k}$ lb
$\mathbf{M}_A = 14,440\mathbf{i} - 64,300\mathbf{j} + 10\mathbf{k}$ lb-ft
$\mathbf{F}_D = -10\mathbf{i} - 3\mathbf{j} + 2150\mathbf{k}$ lb
$\mathbf{M}_D = 14,500\mathbf{i} - 31,125\mathbf{j} + 55\mathbf{k}$ lb-ft

40 $A_x = 333$ lb; $A_y = -500$ lb; $A_z = 300$ lb; $T_D = 1000$ lb
$T_B = 500$ lb; $F_{BF} = -233$ lb;
$\mathbf{F}_E = -333\mathbf{i} + 500\mathbf{j}$; $\mathbf{M}_E = 2250\mathbf{i} - 5000\mathbf{k}$

41 $A_x = 0$; $A_y = -600$ lb; $A_z = 200$ lb
$B_y = -100$ lb; $B_z = 0$; $C_y = 600$ lb

42 $B_y = 3000$ lb; $A_y = 0$
$A_z = B_z = 26,500$ lb
$C_z = 17,000$ lb (front wheel)

43 $T_{AC} = 428$ lb; $T_{AB} = 250$ lb; $F_x = 125$ lb; $F_y = -88$ lb
$F_z = 150$ lb; $C_x = 0$; $C_y = 12,000$ lb-ft; $C_z = -7040$ lb-ft

44 17.55 tons (left supports)
$B_z = 24.9$ tons; $M_D = 70$ ton-ft clockwise

46 $P = 33.7$ lb; $A_z = 25$ lb; $A_y = 7.51$ lb
$B_x = 12.5$ lb; $B_y = 26.31$ lb; $B_z = 0$

47 14,450 lb

48 16,720 lb

49 11,400 lb

50 $T_{BC} = 651$ lb; $T_{DB} = 0$; $F_{AB} = 593$ lb

51 $K = 0$
$A = 433i - 533j$
$H = -433i + 433j - 520k$
$E = 240k$

52 $F_D = 141.5$ lb; $A_y = -105.8$ lb; $A_z = 73.4$ lb
$B_x = 83.1$ lb; $B_y = 58.9$ lb; $B_z = 73.6$ lb

53 $AE = 312$ lb; $FE = 527$ lb; $DE = 312$ lb

54 $A = 310$ lb; $D = 1032$ lb; $C_y = -600$ lb
$C_x = 0$; $C_z = -320$ lb

55 65,500 lb

56 $F_x = A_x = 806$ lb; $F_y = A_y = 500$ lb

57 $A = -64.5j - 217k$ lb
$B = -172.5j - 89k$ lb

58 $A = 260$ tons; $B = 276$ tons; $C = 259$ tons
$D_x = 0$; $D_y = 250$ tons
$E_x = F_x = 0$
$E_y = 15.84$ tons; $F_y = 9.17$ tons

59 $C = \dfrac{Wl^2 \sin \phi}{4\left[\dfrac{l^2}{2}(1 - \cos \phi) + h^2\right]^{1/2}}$

60 $\tan \beta = \dfrac{\dfrac{W_2}{\tan \alpha_1} - \dfrac{W_1}{\tan \alpha_2}}{(W_1 + W_2)}$

61 $P = 4990$ lb; $A_x = 2086$ lb; $A_y = -2140$ lb
$B_x = 403$ lb; $B_y = -2180$ lb; $A_z + B_z = 30$ lb

62 $A_x = 100$ lb; $A_y = 0$; $A_z = -50$ lb
$C_x = 1100$ lb-ft; $C_y = -650$ lb-ft; $C_z = -2200$ lb-ft

63 CD is compressed 600 lb. Force in spring is 1620 lb

64 $A_x = -28$ lb; $A_y = -98$ lb; $B_x = 70$ lb; $B_y = 0$

65 $A = 10,630$ lb; $C = D = 6530$ lb
$G = H = 7510$ lb; $F = 10,620$ lb

66 3 ft; 50-lb vertical forces at supports.

67 $G_x = 750$ lb; $G_y = 400$ lb

68 $F_{CD} = 1700$ lb; $F_{EF} = 4940$ lb

69 $F_1 = W$; $C_2 = -2W$; $F_3 = \frac{1}{22}C_1$

70 794 lb

71 282,000 lb-ft

72 EC = 866 C; ED = 300 T; DC = 612 T; DB = 1433 T
   BC = 0; BA = 1433 T; AC = 2026 lb C
   C = compression; T = tension.

73 CD = −3610 lb C; ED = 2000 lb T; EF = 2000 lb T
   CE = 2000 lb T; AF = 4820 lb T
   AB = 2670 lb C; FC = 1200 lb T

74 AD = 968 lb T; AB = 1985 lb T
   AC = 1740 lb C; BC = 2600 lb C

75 AB = 5.83 K C; AH = 4.66 K T; BH = 1 K T
   HG = 4.66 K T; BG = 1.67 K C
   BC = 4.16 K C; CG = 5 K T

76 AF = −47,600 lb C; AB = 33,677 lb T; FG = −33,677 lb C
   FB = 32,650 lb T; BG = −16,000 lb C
   BC = 45,000 lb T; CD = 45,000 lb T; CG = 20,900 lb T
   GH = FG = 33,677 lb C; GD = BG = 16,000 lb C
   HE = AF = 47,600 lb C; HD = FB = 32,650 lb T
   DE = AB = 33,677 lb T

77 FH = EH = 4.72 K T; FE = 2.5 K T; FC = 0

78 EF = 0.5 K C; FI = 0.558 K C; DH = 1.12 K T

79 (a) DC = 1200 lb T
   (b) DE = 0

80 CB = 250 lb C; BE = 500 lb C

81 15 K T

82 EG = 916,000 lb C
   FH = 836,000 lb T
   IJ = 144,200 lb T

83 707 lb C

84 CB = 902 lb T; EB = 1490 lb C; AB = 640 lb T

85 BC = AC = AD = 0; BD = 1158 lb T; CD = 583 lb C

86 CD = 1500 lb T; CA = CE = 1060 lb C
   ED = 1030 lb C; AE = 1250 lb T

87 FD = 0; EF = 0; AC = 14.1 K; CE = 0
   CD = 14 K; AE = 0; DE = 2 K; AD = 245 K

88 CD = 0; DF = 0; DE = 1000 lb C; BC = 0
   CE = CH = 0; BH = 0; BF = 3650 lb C; BA = 3470 lb T

Supporting forces

A = 666**i** − 666**j** − 3340**k** lb; **H** = **0**; **E** = 1000**k**
F = −666**i** − 1332**j** + 3330**k** lb

89 16.7°

90 (a) 150.5 lb
   (b) 139.5 lb

92 128.1 lb

94 14.7°

95 19,350 ft

96 57.5°

97 $x = a \cos\left[\tan^{-1}\dfrac{\dfrac{aW}{1+\mu_s^2} - \dfrac{aW}{2}}{\dfrac{a\mu_s W}{1+\mu_s^2} + \dfrac{bW}{2}}\right]$

98 938 lb

99 50 lb

100 $\theta = 31°; T = 292$ lb

102 4.76 ft

103 15 ton-ft

104 0.225

105 15.8 in.

106 1319 lb

107 1500 lb

108 $N = 1667$ lb; $f = 100$ lb

109 200 lb

110 (a) 0.144 ft

(b) $\omega = 6.33$ rad/sec

111 $f = 115.5$ lb; $\mu = 0.578$

112 6.8 ft

113 $C = \dfrac{0.483a + 0.1295d}{0.966\mu_s + 0.259} - d$

114 93 lb-ft

115 143.5 lb-ft

116 $P = 226$ lb; $T = 380$ lb

118 374.6 lb

119 115.5 lb

122 0.99 in.

123 $P = 14.9$ lb

125 $P = 210$ lb; $AB = 150$ lb

126 0.62

127 (a) $W_2 = \dfrac{W_1 l}{C}(\mu_1 \cos\alpha - \sin\alpha)\sin\alpha\cos\alpha$

(b) $\mu_2 = \dfrac{\dfrac{1}{2d}(\mu_1\cos\alpha - \sin\alpha)}{\left(\dfrac{\mu_1}{C} + \dfrac{1}{2d}\right)\cos\alpha + \left(\dfrac{\mu_1}{2d} - \dfrac{1}{C}\right)\sin\alpha}$

129 $\tan \alpha = \dfrac{\mu(\tan^2 \beta + 1)}{\mu^2 - \tan^2 \beta}$

$N_B = \dfrac{W}{2(1 + \mu^2)} \dfrac{(\sin \beta - \mu \cos \beta)}{(\sin \beta \cos \beta)}$

$N_a = \dfrac{W}{2(1 + \mu^2)} \dfrac{(\sin \beta + \mu \cos \beta)}{(\sin \beta \cos \beta)}$

130 1.5 ton-ft

131 0.249 max

133 103.5 lb

134 $P = 0.036$ lb; $\alpha = 0.20°$

135 $P = 9.7$ lb; $0.307\mathbf{i} + 0.900\mathbf{j} - 0.376\mathbf{k}$

136 $T = \dfrac{\mu P}{3} \left(\dfrac{D_2^3 - D_1^3}{D_2^2 - D_1^2} \right)$

137 43.6 in.-lb

138 $0.273\,P$ lb-in.

139 1050 lb-in.

140 1055 lb-in.

141 319 lb

142 $2\mu_d(p_0 + p_0')br^2 \sin \dfrac{\theta}{2}$

APPENDIX

1 $F = 65.7$ lb; $\alpha = 32.5°$

2 $F = 38.5$ lb; $\phi = 113.4°$

3 $F = 15.15$ lb; $\alpha = 15.85°$

4 $F = 846$ lb; $\alpha = 17.7°$

5 $|\mathbf{A}| = 220$ lb; $|\mathbf{F}| = 683$ lb

6 66.6 lb

7 $\alpha = 55.9°$; $\beta = 45.5°$

8 $B = 17.3$ lb; $\alpha = 60°$

9 $D = 22.9$ lb

10 $F = 14.10$ lb

11 $F_2 = 516$ lb; $F_1 = 732$ lb

12 $F_s = 36.4$ lb; $F_n = 81.5$ lb

13 $0.627|\mathbf{F}|$; $0.592|\mathbf{F}|$

14 111.6 lbf; 55.9 lbf

15 $\alpha = 90.6°$; 707 lb

17 500 lb

18 (a) 37.4 lb

(b) $l = 0.267$; $m = 0.535$; $n = -0.804$

20 $\mathbf{F} = 70\mathbf{i} - 20\mathbf{j} + 68.6\mathbf{k}$

21 $F_x = 182.2$ lb; $F_y = 912$ lb; $F_z = 365$ lb

22 $F_x = 18.6$ lb; $F_y = 31$ lb; $F_z = 93$ lb

$l = 0.186$; $m = 0.31$; $n = 0.93$

23 729 lb

24 $F_x = 35.4$ lb; $F_y = 35.4$ lb; $F_z = 86.6$ lb

$l = 0.354$; $m = 0.354$; $n = 0.866$

25 209 lb; $l = 0.48$; $m = 0.866$; $n = 0.143$

26 $\mathbf{F} = 25.7\mathbf{i} + 24.7\mathbf{j} + 16\mathbf{k}$

27 $9.7\mathbf{i} - 10.7\mathbf{j} + 16\mathbf{k}$

28 $\mathbf{A} = \pm 5\sqrt{2}\mathbf{i} \mp 5\sqrt{2}\mathbf{j}$

29 $\mathbf{F} = 59.8\mathbf{i} - 71.8\mathbf{j} + 35.8\mathbf{k}$

30 $\hat{\mathbf{r}} = 0.515\mathbf{i} + 0.686\mathbf{j} - 0.515\mathbf{k}$

31 $\mathbf{j} = 0.707\boldsymbol{\epsilon}_r + 0.707\boldsymbol{\epsilon}_\phi$

$\mathbf{F} = 333\mathbf{i} + 667\mathbf{j} + 667\mathbf{k}$ lb

$\mathbf{F} = 708\boldsymbol{\epsilon}_r + 236\boldsymbol{\epsilon}_\phi + 667\boldsymbol{\epsilon}_z$ lb

33 42.1 lb

34 $\mathbf{A} \cdot \mathbf{B} = -164$; $\cos(\mathbf{A}, \mathbf{B}) = -0.465$; -10.5

35 (a) $-6\mathbf{i} + 48\mathbf{j} + 10\mathbf{k}$

(b) $380\mathbf{j}$

36 48 ft-lb

37 -44 ft-lb

39 (a) 75 ft²; $\theta = 95.4°$

(b) 625 ft; $\theta = 21.6°$

40 -29.1 lb

41 $16q$ joules

42 $l = 0.816$; $m = 0.408$; $n = 0.408$; $x = .6$

43 719.25

44 43.4

45 $\mathbf{A} \times \mathbf{B} = -36\mathbf{k} - 18\mathbf{j}$

$\mathbf{B} \times \mathbf{A} = 36\mathbf{k} + 18\mathbf{j}$

$\mathbf{A} \times \mathbf{B} = 40.3$

$l = 0$; $m = -0.447$; $n = -0.894$

48 $\mathbf{A} \times \mathbf{B} = \pm 62.4\mathbf{k}$

49 $\mathbf{A} \times \mathbf{B} = 18\mathbf{i} + 20\mathbf{j} - 42\mathbf{k}$

$\mathbf{A} \cdot \mathbf{B} = 47$

50 $-80\mathbf{i} + 40\mathbf{j} + 50\mathbf{k}$ ft²

51 $-156\mathbf{i} + 94\mathbf{j} + 28\mathbf{k}$ ft²/sec

52 $0.805\mathbf{i} + 0.466\mathbf{j} + 0.372\mathbf{k}$

53 91.5°

54 $\theta = 25°$

55 Area vector of $ADE = -800\mathbf{j} + 500\mathbf{k}$ ft^2
Area vector of $BCE = 800\mathbf{j} - 100\mathbf{k}$ ft^2
Area vector of $ABE = 1600\mathbf{i} + 300\mathbf{k}$ ft^2
Area vector of $CDE = -1600\mathbf{i} + 100\mathbf{k}$ ft^2
Projection of $ADE = 640$ ft^2
Projection of $ABE = 960$ ft^2

58 (a) $-43\mathbf{i} + 49\mathbf{j} + 2\mathbf{k}$
(b) -136
(c) -136

59 2600 ft^3

60 -29.6

61 251 ft^2

64 $-390\mathbf{i} - 1446\mathbf{j} - 1680\mathbf{k}$

65 (a) $-13\mathbf{i} + 49\mathbf{j} + 12\mathbf{k}$
(b) $368\mathbf{i} + 112\mathbf{j} + 160\mathbf{k}$
(c) $48\mathbf{i} - 80\mathbf{j} + 160\mathbf{k}$
(d) 164

66 $-83{,}000\mathbf{i} - 308{,}000\mathbf{j} - 358{,}000\mathbf{k}$

INDEX